Chicago Lectures in Mathematics Series
J. Peter May, Robert J. Zimmer, Spencer J. Bloch, and
Norman R. Lebovitz, editors

The Theory of Sheaves, by Richard G. Swan (1964)
Topics in Ring Theory, by I. N. Herstein (1969)
Fields and Rings, by Irving Kaplansky (1969; 2d ed. 1972)
Infinite Abelian Group Theory, by Philip A. Griffith (1970)
Topics in Operator Theory, by Richard Beals (1971)
Lie Algebras and Locally Compact Groups, by Irving Kaplansky (1971)
Several Complex Variables, by Raghavan Narasimhan (1971)
Torsion-Free Modules, by Eben Matlis (1973)
The Theory of Bernoulli Shifts, by Paul C. Shields (1973)
Stable Homotopy and Generalized Homology, by J. F. Adams (1974)
Commutative Rings, by Irving Kaplansky (1974)
Banach Algebras, by Richard Mosak (1975)
Rings with Involution, by I. N. Herstein (1976)
Theory of Unitary Group Representation, by George W. Mackey (1976)
Infinite-Dimensional Optimization and Convexity, by Ivar Ekeland and
Thomas Turnbull (1963)
Commutative Semigroup Rings, by Robert Gilmer (1984)
Navier-Stokes Equations, by Peter Constantin and Ciprian Foias (1988)
Essential Results of Functional Analysis, by Robert J. Zimmer (1990)
Fuchsian Groups, by Svetlana Katok (1992)

SIMPLICIAL OBJECTS IN ALGEBRAIC TOPOLOGY

J. PETER MAY

The University of Chicago Press
Chicago and London

The University of Chicago Press, Chicago 60637
The University of Chicago Press, Ltd., London

 Published 1967
Midway reprint 1982
Chicago Lectures in Mathematics 1992
Printed in the United States of America

LCN: 82-51078
ISBN: 0-226-51181-2

∞ The paper used in this publication meets the minimum requirements of the American National Standard for Information Sciences—Permanence of Paper for Printed Library Materials, ANSI Z39.48-1984.

INTRODUCTION

Algebraic topology could perhaps be characterized as the study of those functors from the category of topological spaces to that of groups which are invariants of homotopy type. From this point of view, the category of topological spaces which are of the homotopy type of a CW-complex is equivalent to the category of simplicial sets (in the literature: complete semi-simplicial complexes) which satisfy the extension condition.

The object of these notes is an investigation of the category of simplicial sets. There are various technical advantages to this category. These mainly arise from the fact that, for every $n \geq 1$ and Abelian group π, there exists an explicit canonical simplicial Abelian group $K(\pi, n)$ which satisfies $\pi_n(K(\pi, n)) = \pi$ and $\pi_i(K(\pi, n)) = 0, \ i \neq n$. These objects are fundamental to the construction of Postnikov systems and to the study of cohomology operations.

We will first develop the definitions and elementary properties of simplicial objects and then begin the study of simplicial fiber spaces and Postnikov systems. After demonstrating the "equivalence of categories" described above, we will study fiber spaces and fiber bundles in some detail, making use of the concept of "twisted Cartesian product."

We will then study $K(\pi, n)$'s and introduce the k-invariants of Postnikov systems. This will be done by studying simplicial fibre bundles with fibre a $K(\pi, n)$ rather than by use of obstruction theory. We will conclude by developing the Serre spectral sequence by means of Brown's theorem comparing twisted Cartesian products and twisted tensor products.

Most of the material here is scattered through the literature. A bibliographical note at the end of each chapter will give references.

CONTENTS

SIMPLICIAL OBJECTS IN ALGEBRAIC TOPOLOGY

CHAPTER 1
SIMPLICIAL OBJECTS AND HOMOTOPY

§1. Definitions and examples

We introduce the concept of simplicial set and give several examples here. A categorical definition will be given in the next section.

DEFINITIONS 1.1: A simplicial set K is a graded set indexed on the non-negative integers together with maps $\partial_i\colon K_q \to K_{q-1}$ and $s_i\colon K_q \to K_{q+1}$, $0 \leq i \leq q$, which satisfy the following identities:

$$\begin{array}{lrcll}
\text{(i)} & \partial_i \partial_j & = & \partial_{j-1} \partial_i & \text{if } i < j, \\
\text{(ii)} & s_i s_j & = & s_{j+1} s_i & \text{if } i \leq j, \\
\text{(iii)} & \partial_i s_j & = & s_{j-1} \partial_i & \text{if } i < j, \\
 & \partial_j s_j & = & \text{identity} = \partial_{j+1} s_j, & \\
 & \partial_i s_j & = & s_j \partial_{i-1} & \text{if } i > j+1
\end{array}$$

The elements of K_q are called q-simplices. The ∂_i and s_i are called face and degeneracy operators. A simplex x is degenerate if $x = s_i y$ for some simplex y and degeneracy operator s_i; otherwise x is non-degenerate.

DEFINITION 1.2: A simplicial map $f\colon K \to L$ is a map of degree zero of graded sets which commutes with the face and degeneracy operators; that is, f consists of $f_q\colon K_q \to L_q$ and

$$f_q \partial_i = \partial_i f_{q+1},$$

$$f_q s_i = s_i f_{q-1}.$$

DEFINITION 1.3: A simplicial set K is said to satisfy the extension condition if for every collection of $n+1$ n-simplices $x_0, x_1, \ldots, x_{k-1}, x_{k+1}, \ldots, x_{n+1}$ which satisfy the compatibility condition $\partial_i x_j = \partial_{j-1} x_i$, $i < j$, $i \neq k$, $j \neq k$, there exists an $(n+1)$-simplex x such that $\partial_i x = x_i$ for $i \neq k$.

EXAMPLE 1.4: We recall that a simplicial complex K is a set of finite subsets, called simplices, of a given set $\overline{K}$ subject to the condition that every non-empty subset of an element of K is itself an element of K. A simplicial set $\tilde{K}$ arises from K in the following manner. An n-simplex of $\tilde{K}$ is a sequence $(a_0, \ldots a_n)$ of elements of $\overline{K}$ such that the set $\{a_0, \ldots, a_n\}$ is an m-simplex of K for some $m \leq n$. The face and degeneracy operators of $\tilde{K}$ are defined by:

$$\partial_i(a_0, \ldots, a_n) = (a_0, \ldots, a_{i-1}, a_{i+1}, \ldots, a_n)$$

and

$$s_i(a_0, \ldots, a_n) = (a_0, \ldots, a_i, a_i, a_{i+1}, \ldots, a_n) .$$

If the elements of $\overline{K}$ are ordered and we require $\tilde{K}$ to consist of those sequences $(a_0, \ldots, a_n)$ such that $a_0 \leq a_1 \leq \cdots \leq a_n$ and $\{a_0, \ldots, a_n\}$ is an m-simplex of K for some $m \leq n$, then there will be exactly one non-degenerate n-simplex of $\tilde{K}$ for every n-simplex of K.

EXAMPLE 1.5: Let $\Delta_n = \{(t_0, \ldots, t_n) | \, 0 \leq t_i \leq 1, \ \Sigma t_i = 1\} \subset R^{n+1}$. If X is a topological space, a singular n-simplex of X is a continuous function $f\colon \Delta_n \to X$. The graded set $S(X)$, where $S_n(X)$ is the set of singular n-simplices of X, is called the total singular complex of X. $S(X)$ becomes a simplicial set if we define face and degeneracy operators by:

$$(\partial_i f)(t_0, \ldots, t_{n-1}) = f(t_0, \ldots, t_{i-1}, 0, t_i, \ldots, t_{n-1})$$

and

$$(s_i f)(t_0, \ldots, t_{n+1}) = f(t_0, \ldots, t_{i-1}, t_i + t_{i+1}, t_{i+2}, \ldots, t_{n+1}) .$$

The following elementary fact will later be used to show that $S(X)$ determines the homotopy groups of X.

LEMMA 1.5: $S(X)$ satisfies the extension condition.

Proof: Since the union of any $n+1$ faces of Δ_{n+1} is a retract of Δ_{n+1}, any continuous function defined on such a union can be extended to Δ_{n+1}.

CONVENTIONS 1.6: The word "complex" (unmodified) will always mean simplicial set. A complex which satisfies the extension condition will be called a Kan complex.

§ 2. Simplicial objects in categories; homology

Recall that a category $\mathcal{C}$ is a class of objects together with a family of disjoint sets $\mathrm{Hom}(A, B)$, one for each pair of objects, a function $\mathrm{Hom}(B, C) \times \mathrm{Hom}(A, B) \to \mathrm{Hom}(A, C)$, $\alpha \times \beta \to \alpha\beta$, and an element $1_A \in \mathrm{Hom}(A, A)$, all subject to the conditions $\alpha(\beta\gamma) = (\alpha\beta)\gamma$ whenever either is defined and $\alpha 1_A = \alpha = 1_B \alpha$, $\alpha \in \mathrm{Hom}(A, B)$. The elements of $\mathrm{Hom}(A, B)$ are morphisms with domain A and range B. The opposite category $\mathcal{C}^*$ of a category $\mathcal{C}$ has an object A^* for each object A of $\mathcal{C}$ and a morphism $\alpha^* \in \mathrm{Hom}(B^*, A^*)$ for each morphism $\alpha \in \mathrm{Hom}(A, B)$; $\alpha^*\beta^*$ is defined and equal to $(\beta\alpha)^*$ whenever $\beta\alpha$ is defined.

A covariant (resp., contravariant) functor $F\colon \mathcal{C} \to \mathcal{D}$ is a correspondence which assigns to each object $A \in \mathcal{C}$ an object $F(A) \in \mathcal{D}$ and to each morphism $\alpha \in \mathrm{Hom}(A, B)$ a morphism $F(\alpha) \in \mathrm{Hom}(F(A), F(B))$ (resp., $F(\alpha) \in \mathrm{Hom}(F(B), F(A))$) subject to the conditions $F(1_A) = 1_{F(A)}$, $A \in \mathcal{C}$, and $F(\alpha\beta) = F(\alpha)F(\beta)$ (resp., $F(\alpha\beta) = F(\beta)F(\alpha)$) whenever $\alpha\beta$ is defined in $\mathcal{C}$. If $T\colon \mathcal{C} \to \mathcal{C}^*$ is defined by $T(A) = A^*$ and $T(\alpha) = \alpha^*$, then T is a contravariant functor; any contravariant functor $F\colon \mathcal{C} \to \mathcal{D}$ may be considered as the covariant functor $TF\colon \mathcal{C} \to \mathcal{D}^*$ or $FT\colon \mathcal{C}^* \to \mathcal{D}$. If F and G are covariant (resp., contravariant) functors $\mathcal{C} \to \mathcal{D}$, a natural transformation $\lambda\colon F \to G$ is a function which assigns to each object A of $\mathcal{C}$ a morphism $\lambda(A) \in \mathrm{Hom}(F(A), G(A))$ subject to the condition that if

$a \in \mathrm{Hom}(A, B)$, then $G(a)\lambda(A) = \lambda(B)F(a)$ (resp., $G(a)\lambda(\beta) = \lambda(A)F(a))$.

Now we define a category Δ^* as follows. The objects Δ_n of Δ^* are sequences of integers, $\Delta_n = (0, 1, \dots, n)$, $n \geq 0$. The morphisms of Δ^* are the monotonic maps $\mu\colon \Delta_n \to \Delta_m$, that is, the maps μ such that $\mu(i) \leq \mu(j)$ if $i < j$. Define morphisms $\delta_i\colon \Delta_{n-1} \to \Delta_n$ and $\sigma_i\colon \Delta_{n+1} \to \Delta_n$, $0 \leq i \leq n$, by

(1) $\qquad \delta_i(j) = j \text{ if } j < i; \qquad \delta_i(j) = j+1 \text{ if } j \geq i,$

(2) $\qquad \sigma_i(j) = j \text{ if } j \leq i; \qquad \sigma_i(j) = j-1 \text{ if } j > i.$

Let $\mu \in \mathrm{Hom}(\Delta_n, \Delta_m)$, μ not an identity. Suppose $i_1, \dots, i_s$, in reverse order, are the elements of Δ_m not in $\mu(\Delta_n)$ and $j_1, \dots, j_t$, in order, are the elements of Δ_n such that $\mu(j) = \mu(j+1)$. Then:

(3) $$\mu = \delta_{i_1} \dots \delta_{i_s} \sigma_{j_1} \dots \sigma_{j_t}, \text{ where } 0 \leq i_s < \dots < i_1 \leq m,$$
$$0 \leq j_1 < \dots < j_t < n, \text{ and } n - t + s = m.$$

Further, the factorization of μ in the form (3) is unique. Having defined Δ^*, we can formulate

DEFINITIONS 2.1: A simplicial object in a category $\mathcal{C}$ is a contravariant functor $F\colon \Delta^* \to \mathcal{C}$. Such functors form the objects of a category $\mathcal{C}^s$, the morphisms of which are all natural transformations of such functors. If $F \in \mathcal{C}^s$, the elements of $F(\Delta_n)$ are called n-simplices, and the maps $\partial_i = F(\delta_i)$ and $s_i = F(\sigma_i)$ satisfy (i)–(iii) of Definitions 1.1. Any simplicial set K determines a contravariant functor $F\colon \Delta^* \to \mathcal{C}$, where $\mathcal{C}$ is the category of sets, by the rules $F(\Delta_n) = K_n$ and

$$F(\mu) = s_{j_t} \dots s_{j_1} \partial_{i_s} \dots \partial_{i_1},$$

where μ is a morphism of Δ^* expressed in the form (3). Thus a simplicial set may be uniquely identified with a simplicial object in the category of sets. Analogously, we will speak of simplicial groups, simplicial modules, and so forth, depending on the choice of the category $\mathcal{C}$.

REMARKS 2.2: Let $\Delta = \Delta^{**}$ denote the opposite category of Δ^*, $T: \Delta^* \to \Delta$ the contravariant functor defined above. The category $\mathcal{C}^s$ could equally well be defined as that of covariant functors from Δ to $\mathcal{C}$.

Now suppose that $F: \mathcal{C} \to \mathcal{D}$ is a covariant functor. By composition, F induces a covariant functor $F^s: \mathcal{C}^s \to \mathcal{D}^s$. In particular, suppose that $\mathcal{C}$ is the category of sets, $\mathcal{D}$ that of Abelian groups, and for $A \in \mathcal{C}$, $F(A)$ is the free Abelian group generated by A. Then if $K \in \mathcal{C}^s$, $F^s(K)$ may be given a structure of chain complex with differential d defined on $F^s(K)_n = F^s(K_n)$ by

$$d = \sum_{i=0}^{n} (-1)^i \partial_i \quad .$$

We denote this chain complex by $C(K)$. If G is an Abelian group, we define the homology and cohomology of K with coefficients in G by

$$H_*(K; \mathrm{G}) = H(C(K) \otimes \mathrm{G}) \text{ and } H^*(K; \mathrm{G}) = H(\mathrm{Hom}(C(K), \mathrm{G})) \; .$$

In case $K = S(X)$, these are, of course, the singular homology and cohomology groups of the space X.

§3. Homotopy of Kan complexes.

DEFINITION 3.1: Let K be a complex. Two n-simplices x and x' of K are homotopic, written $x \sim x'$, if $\partial_i x = \partial_i x'$, $0 \leq i \leq n$, and there exists a simplex $y \in K_{n+1}$ such that $\partial_n y = x$, $\partial_{n+1} y = x'$, and $\partial_i y = s_{n-1} \partial_i x = s_{n-1} \partial_i x'$, $0 \leq i < n$. The simplex y is called a homotopy from x to x'.

PROPOSITION 3.2: If K is a Kan complex, then $\sim$ is an equivalence relation on the n-simplices of K, $n \geq 0$.

Proof: The relation $\sim$ is reflexive since

$$\partial_n s_n x = x = \partial_{n+1} s_n x$$

and $\partial_i s_n x = s_{n-1} \partial_i x$, $0 \leq i < n$. Suppose $x \sim x'$ and $x \sim x''$.

We must prove $x' \sim x''$. Let y' satisfy

$$\partial_n y' = x, \quad \partial_{n+1} y' = x', \text{ and } \partial_i y' = s_{n-1}\partial_i x', \quad i < n.$$

Let y'' satisfy

$$\partial_n y'' = x, \quad \partial_{n+1} y'' = x'', \text{ and } \partial_i y'' = s_{n-1}\partial_i x', \quad i < n.$$

Then the $n+2$ $(n+1)$-simplices

$$\partial_0 s_n s_n x', \ldots, \partial_{n-1} s_n s_n x', y', y''$$

are seen to satisfy the compatibility condition. Therefore there exists an $(n+2)$-simplex z such that $\partial_i z = \partial_i s_n s_n x'$, $0 \leq i < n$, $\partial_n z = y'$, and $\partial_{n+1} z = y''$. It follows that

$$\partial_i \partial_{n+2} z = s_{n-1} \partial_i x', \quad 0 \leq i < n,$$

$\partial_n \partial_{n+2} z = x'$, and $\partial_{n+1} \partial_{n+2} z = x''$, hence $x' \sim x''$.

DEFINITION 3.3: Let L be a subcomplex of K. Two n-simplices x and x', $n > 0$, are homotopic relative to L, written $x \sim x'$ rel L, if $\partial_i x = \partial_i x'$, $1 \leq i \leq n$, if $\partial_0 x \sim \partial_0 x'$ in L, and if there exists a homotopy y from $\partial_0 x$ to $\partial_0 x'$ in L and a simplex $w \in K_{n+1}$ such that $\partial_0 w = y$, $\partial_n w = x$, $\partial_{n+1} w = x'$, and $\partial_i w = s_{n-1} \partial_i x = s_{n-1} \partial_i x'$, $1 \leq i < n$. The simplex w is called a relative homotopy from x to x'.

PROPOSITION 3.4: If L is a sub Kan complex of the Kan complex K, then $\sim$ rel L is an equivalence relation on the n-simplices of K, $n \geq 1$.

Proof: The relation $\sim$ rel L is reflexive since if $\partial_0 x \in L$, then $s_{n-1} \partial_0 x$ is a homotopy in L from $\partial_0 x$ to $\partial_0 x$, and if $w = s_n x$, then $\partial_i w = s_{n-1} \partial_i x$, $0 \leq i < n$, and

$$\partial_n w = x = \partial_{n+1} w.$$

Suppose $x \sim x'$ rel L and $x \sim x''$ rel L. We must prove that $x' \sim x''$ rel L. Let y' and y'' be homotopies in L from $\partial_0 x$ to $\partial_0 x'$ and from $\partial_0 x$ to $\partial_0 x''$, and suppose w' and w'' are relative homotopies from x to x' and from x to x'' which satisfy

$\partial_0 w' = y'$ and $\partial_0 w'' = y''$. As in the proof of Proposition 3.2, we may choose $z \in L_{n+1}$ such that

$$\partial_i z = \partial_i s_{n-1} s_{n-1} \partial_0 x', \quad 0 \leq i < n-1,$$

$$\partial_{n-1} z = y' \quad \text{and} \quad \partial_n z = y''.$$

Then $y = \partial_{n+1} z$ is a homotopy in L from $\partial_0 x'$ to $\partial_0 x''$. Now it is easy to see that the $n+2$ $(n+1)$-simplices

$$z, \partial_1 s_n s_n x', \ldots, \partial_{n-1} s_n s_n x', w', w''$$

satisfy the compatibility condition so that there exists $v \in K_{n+2}$ such that $\partial_i v = \partial_i s_n s_n x'$, $1 \leq i < n$, $\partial_0 v = z$, $\partial_n v = w'$, and $\partial_{n+1} v = w''$. Let $w = \partial_{n+2} v$. Then $\partial_i w = s_{n-1} \partial_i x'$, $1 \leq i < n$, $\partial_0 w = y$, $\partial_n w = x'$, and $\partial_{n+1} w = x''$.

NOTATIONS 3.5: Let K be a complex, $\phi \in K_0$. ϕ generates a subcomplex of K which has exactly one simplex $s_{n-1} \ldots s_0 \phi$ in each dimension n. We will abuse notation by letting ϕ denote ambiguously either this subcomplex or any of its simplices. We call (K, ϕ) a Kan pair if K is a Kan complex. We call (K, L, ϕ) a Kan triple if $\phi \in L_0$ and L is a sub Kan complex of the Kan complex K. Simplicial maps of pairs and triples are defined in the obvious manner.

DEFINITIONS 3.6: Let (K, ϕ) be a Kan pair. Let $\tilde{K}_n$, $n \geq 0$, denote the set of all $x \in K_n$ which satisfy $\partial_i x = \phi$, $0 \leq i \leq n$. Then we define $\pi_n(K, \phi) = \tilde{K}_n/(\sim)$. $\pi_0(K, \phi)$ is called the set of path components of K. K is connected if $\pi_0(K, \phi) = \phi$ (where we are letting ϕ denote its equivalence class). K is n-connected if $\pi_i(K, \phi) = \phi$, $0 \leq i \leq n$. Let (K, L, ϕ) be a Kan triple. Let $\tilde{K}(L)_n$, $n \geq 1$, denote the set of all $x \in K_n$ which satisfy $\partial_0 x \in L_{n-1}$ and $\partial_i x = \phi$, $1 \leq i \leq n$. Then we define

$$\pi_n(K, L, \phi) = \tilde{K}(L)_n/(\sim \text{rel } L).$$

Note that $\pi_n(K, \phi, \phi) = \pi_n(K, \phi)$, $n \geq 1$. Finally, we define $\partial\colon \pi_n(K, L, \phi) \to \pi_{n-1}(L, \phi)$, $n \geq 1$, by $\partial[x] = [\partial_0 x]$, where $[x]$ denotes the homotopy class of x.

THEOREM 3.7: Let (K, L, ϕ) be a Kan triple. Then the sequence

$$\cdots \to \pi_{n+1}(K, L, \phi) \xrightarrow{\partial} \pi_n(L, \phi) \xrightarrow{i} \pi_n(K, \phi) \xrightarrow{j} \pi_n(K, L, \phi) \to \cdots$$

of sets with distinguished element ϕ is exact, where the maps i and j are induced by inclusion.

Proof: (i) $i\partial = \phi$: Consider $i[\partial_0 x] = i\partial[x]$, $x \in \tilde{K}(L)_{n+1}$. The $n+2$ $(n+1)$-simplices $-, \phi, \cdots, \phi, x$ are compatible, hence we may choose $z \in K_{n+2}$ such that $\partial_i z = \phi$, $1 \le i \le n+1$, $\partial_{n+2} z = x$. Then $\partial_i \partial_0 z = \partial_0 \partial_{i+1} z = \phi$, $0 \le i \le n+1$, and $\partial_{n+1} \partial_0 z = \partial_0 x$, so that $\partial_0 x \sim \phi$ in K.

(ii) Image ∂ = Kernel i: Let $i[y] = \phi$, $y \in \tilde{L}_n$. Then $y \sim \phi$ in K_n, say $\partial_i z = \phi$, $0 \le i \le n$, $\partial_{n+1} z = y$. The $n+2$ $(n+1)$-simplices $z, \phi, \ldots, \phi$ are compatible, hence there exists $w \in K_{n+2}$ with $\partial_0 w = z$, $\partial_i w = \phi$, $1 \le i \le n+1$. $\partial_i \partial_{n+2} w = \phi$, $1 \le i \le n+1$, and $\partial_0 \partial_{n+2} w = y$, so that $\partial[\partial_{n+2} w] = [y]$.

(iii) $\partial j = \phi$: $\partial j[x] = \phi$ since $\partial_0 x = \phi$, $x \in \tilde{K}_n$.

(iv) Image j = Kernel ∂: Let $\partial[x] = [\partial_0 x] = \phi$, $x \in \tilde{K}(L)_n$. There exists $z \in L_n$ such that $\partial_i z = \phi$, $0 \le i \le n$, $\partial_n z = \partial_0 x$. The $n+1$ n-simplices $z, \phi, \ldots, \phi, -, x$ (where $-$ means $k = n$ in the compatibility condition) are compatible, say $\partial_0 y = z$, $\partial_i y = \phi$, $1 \le i \le n-1$, $\partial_{n+1} y = x$. Thus $x \sim \partial_n y$ rel L. Since $\partial_i \partial_n y = \phi$, $0 \le i \le n$, $[x] = j[\partial_n y]$.

(v) $ji = \phi$: Consider $ji[y]$, $y \in \tilde{L}_n$. The $n+1$ n-simplices $-, \phi, \ldots, \phi, y$ are compatible in L, say $z \in L_{n+1}$ satisfies $\partial_i z = \phi$, $1 \le i \le n$, $\partial_{n+1} z = y$. $\partial_i \partial_0 z = \phi, 0 \le i \le n$, hence $\partial_0 z$ is a homotopy between ϕ and $\phi = \partial_0 y$, so z is a relative homotopy between ϕ and y.

(vi) Image i = Kernel j: Let $j[x] = \phi$, $x \in \tilde{K}_n$. Choose $w \in K_{n+1}$ such that $\partial_i w = \phi$, $1 \le i \le n$, $\partial_{n+1} w = x$, and $\partial_0 w = z \in L_n$. The $n+1$ n-simplices $z, \phi, \cdots, \phi, -$

are compatible in L, say $\partial_0 v = z$ and $\partial_i v = \phi$, $1 \leq i \leq$ n. The $n+2$ $(n+1)$-simplices $s_{n-1}z, \phi, \ldots, \phi, v, w, -$ are compatible in K, say $\partial_0 t = s_{n-1}z$, $\partial_i t = \phi$, $1 \leq i \leq n-1$, $\partial_n t = v$, $\partial_{n+1} t = w$. Then $\partial_{n+2} t$ is a homotopy $\partial_{n+1} v \sim x$, hence $[x] = i[\partial_{n+1} v]$.

§ 4. The group structures.

DEFINITIONS 4.1: Let (K, ϕ) be a Kan pair. Write $x \in a$ if x represents a. Suppose $a, \beta \in \pi_n(K, \phi)$, $n \geq 1$, and let $x \in a$, $y \in \beta$. The $n+1$ n-simplices $\phi, \ldots, \phi, x, -, y$ are compatible, say $\partial_i z = \phi$, $0 \leq i < n-1$, $\partial_{n-1} z = x$, $\partial_{n+1} z = y$. We define $a\beta = [\partial_n z]$.

LEMMA 4.2: $a\beta$ is well defined.

Proof: Suppose z' also satisfies $\partial_i z' = \phi$, $0 \leq i < n-1$, $\partial_{n-1} z' = x$, and $\partial_{n+1} z' = y$. By the extension condition, there exists $w \in K_{n+2}$ such that $\partial_i w = \phi$, $0 \leq i \leq n-2$,

$$\partial_{n-1} w = s_n \partial_{n-1} z, \ \partial_{n+1} w = z, \text{ and } \partial_{n+2} w = z'.$$

$\partial_n w$ is then a homotopy from $\partial_n z$ to $\partial_n z'$. Suppose $y \sim y'$, say $\partial_i w = \phi$, $0 \leq i < n$, $\partial_n w = y'$, $\partial_{n+1} w = y$. Choose z' such that $\partial_{n-1} z' = x$, $\partial_{n+1} z' = y'$, and $\partial_i z' = \phi$, $0 \leq i < n-1$. By the extension condition, there exists $u \in K_{n+2}$ such that $\partial_i u = \phi$, $0 \leq i < n-1$,

$$\partial_{n-1} u = s_{n-1} x, \ \partial_n u = z', \text{ and } \partial_{n+2} u = w.$$

Then $\partial_n \partial_{n+1} u = \partial_n z'$, $\partial_{n+1} \partial_{n+1} u = y$, and $\partial_{n-1} \partial_{n+1} u = x$, so the same choice of z may be taken for the two choices of representatives for β. Similarly $a\beta$ is independent of the choice of the representative for a.

PROPOSITION 4.3: With the above multiplication, $\pi_n(K, \phi)$ is a group, $n \geq 1$.

Proof: (i) Divisibility: Let $x \in a$, $y \in \beta$, a and β in $\pi_n(K, \phi)$. By the extension condition, there exists $z \in K_{n+1}$ such that $\partial_i z = \phi$, $0 \le i < n-1$, $\partial_n z = y$, and $\partial_{n+1} z = x$. Then $[\partial_{n-1} z] \alpha = \beta$. Right divisibility is proven similarly.

(ii) Associativity: Let $x \in \alpha$, $y \in \beta$, $z \in \gamma$, where α, β, and γ are in $\pi_n(K, \phi)$. Using the extension condition, choose $w_{n-1}, w_{n+1}, w_{n+2}$ such that $\partial_i w_j = \phi$, $0 \le i < n-1$,

$$\partial_{n-1} w_{n-1} = x, \quad \partial_{n+1} w_{n-1} = y, \quad \partial_{n-1} w_{n+1} = \partial_n w_{n-1},$$

$$\partial_{n+1} w_{n+1} = z, \quad \partial_{n-1} w_{n+2} = y, \quad \text{and} \quad \partial_{n+1} w_{n+2} = z.$$

Applying the extension condition again, choose $u \in K_{n+2}$ such that $\partial_i u = \phi$, $0 \le i < n-1$, and $\partial_i u = w_i$, $i = n-1, n+1$, $n+2$. Then $\partial_{n-1} \partial_n u = x$, $\partial_{n+1} \partial_n u = \partial_n w_{n+2}$, and $\partial_i u = \phi$, $0 \le i < n-1$. Therefore:

$$(\alpha\beta)\gamma = [\partial_n w_{n-1}]\gamma = [\partial_{n-1} w_{n+1}]\gamma = [\partial_n w_{n+1}]$$

$$= [\partial_n \partial_n u] = \alpha[\partial_n w_{n+2}] = \alpha(\beta\gamma) .$$

PROPOSITION 4.4: $\pi_n(K, \phi)$ is Abelian if $n \ge 2$.

Proof: Let $w, x, y, z \in \tilde{K}_n$.

(i) Suppose $v_{n+1} \in K_{n+1}$ satisfies $\partial_i v_{n+1} = \phi$, $0 \le i < n-2$, $\partial_{n+1} v_{n+1} = \phi$, $\partial_{n-2} v_{n+1} = w$, $\partial_{n-1} v_{n+1} = x$, and $\partial_n v_{n+1} = y$. Then $[y][w] = [x]$: Choose $v_{n-1} \in K_{n+1}$ with faces $\partial_i v_{n-1} = \phi$, $0 \le i \le n-2$, $\partial_n v_{n-1} = x$, and

$$\partial_{n+1} v_{n-1} = w ,$$

and let $t = \partial_{n-1} v_{n-1}$. Let $v_i = \phi$, $0 \le i < n-2$, and let $v_{n-2} = s_n w$, $v_{n+2} = s_{n-2} w$. The v_i satisfy the extension condition, say $\partial_i r = v_i$, $i \ne n$. Let $v_n = \partial_n r$.

$$\partial_i v_n = \phi, \quad 0 \le i \le n-2, \quad \partial_{n-1} v_n = t, \quad \partial_n v_n = y,$$

and $\partial_{n+1} v_n = \phi$. Therefore $[t][\phi] = [y]$; but by the choice of v_{n-1}, $[t][w] = [x]$, hence $[y][w] = [x]$.

(ii) Suppose $v_n \in K_{n+1}$ satisfies $\partial_i v_n = \phi$, $0 \le i < n-2$, $\partial_{n-1} v_n = \phi$, $\partial_{n-2} v_n = w$, $\partial_n v_n = y$, and $\partial_{n+1} v_n = z$.

Then $[w][y] = [z]$: Choose $v_{n-1} \in K_{n+1}$ with faces $\partial_i v_{n-1} = \phi$, $0 \leq i < n-2$, $\partial_{n-2} v_{n-1} = w$, and $\partial_{n-1} v_{n-1} = \phi = \partial_{n+1} v_{n-1}$, and let $t = \partial_n v_{n-1}$. Let $v_i = \phi$, $0 \leq i < n-2$, and let

$$v_{n-2} = s_{n-2} w, \quad v_{n+2} = s_n z.$$

The v_i satisfy the extension condition, say $\partial_i r = v_i$, $i \neq n+1$. Let $v_{n+1} = \partial_{n+1} r$. $\partial_i v_{n+1} = \phi$, $0 \leq i \leq n-2$, $\partial_{n-1} v_{n+1} = t$, $\partial_n v_{n+1} = y$, and $\partial_{n+1} v_{n+1} = z$. Therefore $[t][z] = [y]$; but by the choice of v_{n-1} and (i), $[t][w] = [\phi]$, hence $[w][y] = [z]$.

(iii) Suppose $v_{n+2} \in K_{n+1}$ satisfies $\partial_i v_{n+2} = \phi$, $0 \leq i < n-2$, $\partial_{n-2} v_{n+2} = w$, $\partial_{n-1} v_{n+2} = x$, $\partial_n v_{n+2} = y$, and $\partial_{n+1} v_{n+2} = z$. Then $[w]^{-1}[x][z] = [y]$: Choose $v_{n-2} \in K_{n+1}$ with faces $\partial_i v_{n-2} = \phi$, $i \neq n-2, n+1$, $\partial_{n+1} v_{n-2} = w$; let $t = \partial_{n-2} v_{n-2}$. Choose $v_{n-1} \in K_{n+1}$ with faces $\partial_i v_{n-1} = \phi$ unless $i = n-2$, n, or $n+1$, $\partial_{n-2} v_{n-1} = t$, $\partial_{n+1} v_{n-1} = x$; let $u = \partial_n v_{n-1}$. Let $v_i = \phi$, $0 \leq i < n-2$, $v_n = s_n y$. The v_i satisfy the extension condition, say $\partial_i r = v_i$, $i \neq n+1$. Let $v_{n+1} = \partial_{n+1} r$. By (ii), $[t] = [w]$ and $[t][u] = [x]$. $\partial_i v_{n+1} = \phi$, $0 \leq i \leq n-2$, $\partial_{n-1} v_{n+1} = u$, $\partial_n v_{n+1} = y$, and $\partial_{n+1} v_{n+1} = z$. Therefore $[u][z] = [y]$. Combining, we find $[w]^{-1}[x][z] = [y]$.

(iv) Set $z = \phi$ in (iii). Then $[w]^{-1}[x] = [y]$. But applying (i) to v_{n+2} of (iii) in this case, we find $[y] = [x][w]^{-1}$. Therefore for any $[x]$ and $[w]$, $[w]^{-1}[x] = [x][w]^{-1}$, and this implies the result.

DEFINITION 4.5: Let (K, L, ϕ) be a Kan triple. Suppose $\alpha, \beta \in \pi_n(K, L, \phi)$, $n \geq 2$, and let $x \in \alpha$, $y \in \beta$. We have $[\partial_0 x][\partial_0 y] = [\partial_{n-1} z]$, where $\partial_i z = \phi$, $0 \leq i \leq n-3$, $\partial_{n-2} z = \partial_0 x$, and $\partial_n z = \partial_0 y$. The $n+1$ n-simplices $z, \phi, \ldots, \phi, x, -, y$ are compatible, say $\partial_i w = \phi$, $1 \leq i \leq n-2$, $\partial_0 w = z$, $\partial_{n-1} w = x$, $\partial_{n+1} w = y$. We define $\alpha\beta = [\partial_n w]$.

LEMMA 4.6: $\alpha\beta$ is well defined.

PROPOSITION 4.7: With the above multiplication, $\pi_n(K, L, \phi)$ is a group, $n \geq 2$.

PROPOSITION 4.8: $\pi_n(K, L, \phi)$ is Abelian if $n \geq 3$.

The proofs of the above statements are analogous to those for the absolute groups and will be omitted.

Now maps of Kan pairs or triples induce homomorphisms of homotopy groups in the obvious manner. Further, ∂ is a homomorphism of groups by construction. Thus we find:

PROPOSITION 4.9: π_n is a functor from the category of Kan pairs (resp., triples) to that of groups, $n \geq 1$ (resp., $n \geq 2$), and to the category of sets, $n = 0$ (resp., $n = 1$). ∂ is a natural transformation of functors, $n \geq 1$. In particular, the exact sequence of Theorem 3.7 is functorial and is an exact sequence of groups up to $\pi_1(K, \phi)$.

§5. Homotopy of simplicial maps.

DEFINITIONS 5.1: Let f and g be simplicial maps from a complex K to a complex L. Then f is homotopic to g, written $f \simeq g$, if there exist functions $h_i\colon K_q \to L_{q+1}$, $0 \leq i \leq q$, which satisfy:

(i) $\partial_0 h_0 = f$, $\partial_{q+1} h_q = g$

(ii) $\partial_i h_j = h_{j-1} \partial_i$ if $i < j$

$\partial_{j+1} h_{j+1} = \partial_{j+1} h_j$

$\partial_i h_j = h_j \partial_{i-1}$ if $i > j+1$

(iii) $s_i h_j = h_{j+1} s_i$ if $i \leq j$

$s_i h_j = h_j s_{i-1}$ if $i > j$

h is called a homotopy from f to g, $h\colon f \simeq g$. If K' and L' are subcomplexes of K and L and f and g take K' into L', then

h is a relative homotopy from f to g if h is a homotopy from f to g, if $h(K') \subset L'$, and if $h|K'$ is a homotopy from $f|K'$ to $g|K'$. K' is said to be a deformation retract of K if the identity map of K is homotopic relative to K' to a map from K onto K', which extends the inclusion map $K' \to K$. Two complexes K and L are said to be of the same homotopy type if there exist maps $f\colon K \to L$ and $g\colon L \to K$ which satisfy $fg \simeq 1$ and $gf \simeq 1$.

LEMMA 5.2: Let $f, f'\colon K \to L$ and $g, g'\colon L \to M$. Then:

(i) $f \simeq f$

(ii) If $h\colon f \simeq f'$, then $g \circ h\colon g \circ f \simeq g \circ f'$

(iii) If $h\colon g \simeq g'$, then $h \circ f\colon g \circ f \simeq g' \circ f$.

Proof: If $h_i(x) = s_i f(x)$, then $h\colon f \simeq f$. (ii) and (iii) are clear.

We will prove that homotopy is an equivalence relation on maps from a complex K to a Kan complex L in the next section.

The following proposition implies that homology is an invariant of homotopy type on the category of simplicial sets.

PROPOSITION 5.3: Suppose f and g are maps from K into L. Then if $f \simeq g$, $f_* = g_*\colon H_*(K) \to H_*(L)$.

Proof: Let $h\colon f \simeq g$. Define $s\colon C(K) \to C(L)$ by

$$s(x) = \sum_{i=0}^{q} (-1)^i h_i(x), \qquad x \in K_q.$$

Then it is easily verified that $ds + sd = C(f) - C(g)$, which implies the result.

We now wish to express our definition of homotopy groups in terms of homotopies of maps. We first define the standard simplicial n-simplex $\Delta[n]$.

DEFINITIONS AND NOTATIONS 5.4: For each $n \geq 0$, a simplicial set $\Delta[n]$ is given by the contravariant functor T^n defined by $T^n(\Delta_m) = \operatorname{Hom}(\Delta_m, \Delta_n)$ and $T^n(\mu)(\lambda) = \lambda\mu$ whenever $\lambda\mu$ is defined in the category Δ^*. The face and degeneracy operators are of course $T^n(\delta_i)$ and $T^n(\sigma_i)$. Note that δ_i and σ_i, $0 \leq i \leq n$, induce simplicial maps

$$\delta_i\colon \Delta[n-1] \to \Delta[n] \text{ and } \sigma_i\colon \Delta[n+1] \to \Delta[n].$$

Identifying $\lambda \in \operatorname{Hom}(\Delta_m, \Delta_n)$ with its image $\lambda(\Delta_m)$, an m-simplex of $\Delta[n]$ is a sequence $(a_0,\ldots,a_m)$ of integers a_i, $0 \leq a_0 \leq \cdots \leq a_n \leq n$, and

$$\partial_i(a_0,\ldots,a_m) = (a_0,\ldots,a_{i-1},a_{i+1},\ldots,a_m),$$
$$s_i(a_0,\ldots,a_m) = (a_0,\ldots,a_i,a_i,\ldots,a_m) .$$

We let Δ_n denote $(0,1,\ldots,n) \in \Delta[n]_n$. We denote by $\dot{\Delta}[n]$, $\Delta^0[n]$, and $\Delta^1[n]$ the subcomplexes of $\Delta[n]$ generated by $\{\partial_i\Delta_n \mid 0 \leq i\}$, $\partial_0\Delta_n$, and $\{\partial_i\Delta_n \mid 1 \leq i\}$ respectively.

Now suppose K is a complex, $x \in K_n$. Clearly there is a unique simplicial map $\bar{x}\colon \Delta[n] \to K$ such that $\bar{x}(\Delta_n) = x$. If $x \in \tilde{K}_n$, then $\bar{x}\colon (\Delta[n], \dot{\Delta}[n]) \to (K,\phi)$ and if L is a subcomplex of K and $x \in \tilde{K}(L)_n$, then

$$\bar{x}\colon (\Delta[n], \Delta^0[n], \Delta^1[n]) \to (K, L, \phi) .$$

LEMMA 5.5: Let (K,ϕ) be a Kan pair, and suppose $x, y \in \tilde{K}_n$. Then $x \sim y$ if and only if $\bar{x} \simeq \bar{y}$ rel $\dot{\Delta}[n]$.

Proof: Suppose $x \sim y$, say $z \in K_{n+1}$ satisfies $\partial_i z = \phi$, $0 \leq i < n$, $\partial_n z = x$, and $\partial_{n+1} z = y$. We must define $h\colon \bar{x} \simeq \bar{y}$ rel $\dot{\Delta}[n]$, and it clearly suffices to define h_i on Δ_n. We do this by $h_i(\Delta_n) = s_i x$, $0 \leq i < n$, and $h_n(\Delta_n) = z$. Conversely, suppose $h\colon \bar{x} \simeq \bar{y}$ and let $z_i = h_i(\Delta_n)$, $0 \leq i \leq n$. Then $\partial_i z_j = \phi$ unless $i = j$ or $i = j+1$, and $\partial_0 z_0 = x$, $\partial_i z_i = \partial_i z_{i-1}$ for

$1 \leq i \leq n$, $\partial_{n+1} z_n = y$. We will prove that if $z \in K_{n+1}$, $\partial_i z = \phi$ for $i \neq r$, $r+1$, and $\partial_r z \in \tilde{K}_n$, $\partial_{r+1} z \in \tilde{K}_n$, then $\partial_r z \sim \partial_{r+1} z$. This will prove the result, since it will imply

$$x \sim \partial_1 z_0 = \partial_1 z_1 \sim \partial_2 z_1 = \cdots \sim \partial_n z_{n-1} = \partial_n z_n \sim y.$$

Thus suppose $r < n$, and let $a = \partial_r z$, $b = \partial_{r+1} z$, where $\partial_i z = \phi$ if $i \neq r$, $r+1$. By the extension condition, there exists $w \in K_{n+2}$ such that $\partial_i w = \phi$ if $0 \leq i < r$ and if $r+3 < i \leq n+2$, $\partial_{r+1} w = s_{r+1} b$, $\partial_{r+2} w = z$, and $\partial_{r+3} w = s_r b$. Then $\partial_i \partial_r w = \phi$, $i \neq r+1, r+2$, and $\partial_{r+1} \partial_r w = a$, $\partial_{r+2} \partial_r w = b$. Iterating the process, we find $a \sim b$.

LEMMA 5.6: Let (K, L, ϕ) be a Kan triple, and suppose

$$x, y \in \tilde{K}(L)_n.$$

Then $x \sim y$ rel L if and only if $\bar{x} \simeq \bar{y}$ rel $(\Delta^0[n], \Delta^1[n])$.

The proof is similar to that of Lemma 5.5 and will be omitted. Lemmas 5.5 and 5.6 imply

PROPOSITION 5.7: Let (K, L, ϕ) be a Kan triple. Then:

(i) Homotopy is an equivalence relation on maps

$$(\Delta[n], \dot{\Delta}[n]) \to (K, \phi),$$

and $\pi_n(K, \phi)$ may be identified with the set of equivalence classes of such maps.

(ii) Homotopy is an equivalence relation on maps

$$(\Delta[n], \Delta^0[n], \Delta^1[n]) \to (K, L, \phi),$$

and $\pi_n(K, L, \phi)$ may be identified with the set of equivalence classes of such maps.

Together with (iii) of Lemma 5.2, Proposition 5.7 implies:

PROPOSITION 5.8: If $f, g\colon (K, L, \phi) \to (K', L', \phi')$ and $f \simeq g$

rel (L', ϕ'), then $f_* = g_*\colon \pi_n(K, L, \phi) \to \pi_n(K', L', \phi')$, and similarly for the absolute groups. In particular, the homotopy groups are invariants of homotopy type on the categories of Kan pairs and of Kan triples.

§6. Function complexes

In this section, we reformulate the definition of homotopy of maps and show that maps $K \to L$ are vertices of a complex L^K, the paths of which are homotopies between maps.

DEFINITION 6.1: The Cartesian product $K \times L$ of two complexes K and L is defined as $(K \times L)_q = K_q \times L_q$ with face and degeneracy operators given by

$$\partial_i(x, y) = (\partial_i x, \partial_i y) \quad \text{and} \quad s_i(x, y) = (s_i x, s_i y) \ .$$

NOTATIONS 6.2: Let $I = \Delta[1]$ and $\dot{I} = \dot{\Delta}[1] \cdot (0)$ will denote any simplex of $\Delta^1[1]$, (1) any simplex of $\Delta^0[1]$.

PROPOSITION 6.2: Let f and g be simplicial maps $K \to L$. Then $f \simeq g$ if and only if there exists a simplicial map $F\colon K \times I \to L$ such that $F(x, (0)) = g(x)$ and $F(x, (1)) = f(x)$, where x is any simplex of K.

Proof: Suppose $h\colon f \simeq g$. For $x \in K_q$ define $F(x, (0)) = g(x)$, $F(x, (1)) = f(x)$, *and*

$$F(x, s_{q-1} \cdots s_{i+1} s_{i-1} \cdots s_0 \Delta_1) = \partial_{i+1} h_i(x) \ ,$$

$0 \leq i \leq q-1$. Then it is easily verified that F is a simplicial map. Conversely, given F define

$$h_i(x) = F(s_i x, s_q \cdots s_{i+1} s_{i-1} \cdots s_0 \Delta_1) \ ,$$

$x \in K_q$, $0 \leq i \leq q$. Then $h\colon f \simeq g$.

We will call F a homotopy from g to f, $F\colon g \simeq f$. Similarly we may prove

PROPOSITION 6.3: Let $f, g\colon (K, K') \to (L, L')$. Then $f \simeq g$ rel L if and only if there exists $F\colon K \times I \to L$ such that $F\colon g \simeq f$, $F(K' \times I) \subset L'$, and $F|K' \times I\colon g|K' \simeq f|K'$.

DEFINITION 6.4: The function complex L^K of maps from a complex K to a complex L is defined as follows:

(i) $(L^K)_q = \operatorname{Hom}(K \times \Delta[q], L)$, the set of simplicial maps $F\colon K \times \Delta[q] \to L$

(ii) $\partial_i f = f(1 \times \delta_i)$ and $s_i f = f(1 \times \sigma_i)$, where 1 denotes the identity map of K.

$(L, L')^{(K, K')}$ is similarly defined, with q-simplices the elements of $\operatorname{Hom}((K \times \Delta[q], K' \times \Delta[q]), (L, L'))$.

Now a homotopy $F\colon f \simeq g$ between maps $K \to L$ is just a path (1-simplex) of L^K which starts at $\partial_1 F = f$ and ends at $\partial_0 F = g$. It follows that homotopy is an equivalence relation on maps $K \to L$ whenever L^K is a Kan complex, and similarly for maps of pairs, etc. We will prove that L^K is a Kan complex if L is. To do this, we need a combinatorial description of q-simplices of L^K similar to that of 1-simplices given by the original definition of homotopy of maps.

DEFINITION 6.5: A (p, q)-shuffle is a permutation π of $\{0, \dots, p+q-1\}$ which satisfies $\pi(i) < \pi(j)$ if $0 \le i < j \le p-1$ or if $p \le i < j \le p+q-1$. Let $\mu_i = \pi(i-1)$, $1 \le i \le p$, and let $\nu_j = \pi(j+p-1)$, $1 \le j \le q$. π is determined by μ or ν, and we write $\pi = (\mu, \nu)$.

If $y \in K_p$ is non-degenerate and $\hat{y}$ denotes the subcomplex of K generated by y, then the non-degenerate $(p+q)$-simplices of $\hat{y} \times \Delta[q]$ are $\{(s_{\nu_q} \dots s_{\nu_1} y, s_{\mu_p} \dots s_{\mu_1} \Delta_q) | (\mu, \nu) \text{ is a } (p, q)\text{-shuffle}\}$. This motivates the introduction of shuffles.

Let $i \in \{0,\ldots, p+q\}$. We classify (p, q)-shuffles relative to i as follows. (μ, ν) is of type I if

(i) $i < \mu_1$, or

(ii) $i, i-1 \in \{\nu_1,\ldots,\nu_q\}$, or

(iii) $i = p+q$, $i-1 = \nu_q$.

(μ, ν) if of type II if

(i) $i < \nu_1$, or

(ii) $i, i-1 \in \{\mu_1,\ldots,\mu_p\}$, or

(iii) $i = p+q$, $i-1 = \mu_p$.

(μ, ν) is of type III otherwise; that is (μ, ν) is of type III if

(i) $\max\{\mu_1, \nu_1\} \leq i < p+q$, and either

(ii) $i \in \{\mu_1,\ldots, \mu_p\}$ and $i-1 \in \{\nu_1,\ldots,\nu_q\}$, or

(iii) $i \in \{\nu_1,\ldots,\nu_q\}$ and $i-1 \in \{\mu_1,\ldots,\mu_p\}$.

To each shuffle (μ, ν) and each i we associate a new shuffle $(\bar{\mu}, \bar{\nu})$ and an index r as follows: If (μ, ν) is of type I, $(\bar{\mu}, \bar{\nu})$ is a $(p, q-1)$-shuffle. Let k be that integer such that $i = \nu_k$ in cases (i) and (ii) and let $k = q$ in case (iii). Let $\bar{\nu}_j = \nu_j$ for $j < k$ and let $\bar{\nu}_j = \nu_{j+1} - 1$ for $k \leq j \leq q-1$. This defines $(\bar{\mu}, \bar{\nu})$, and r is that integer such that $\bar{\mu}_j = \mu_j$ for $j \leq r$ and $\bar{\mu}_j = \mu_j - 1$ for $r < j \leq p$.

If (μ, ν) is of type II, $(\bar{\mu}, \bar{\nu})$ is a $(p-1, q)$-shuffle. Let k be that integer such that $i = \mu_k$ in cases (i) and (ii) and let $k = p$ in case (iii). Let $\bar{\mu}_j = \mu_j$ for $j < k$ and let $\bar{\mu}_j = \mu_{j+1} - 1$ for $k \leq j \leq p-1$. This defines $(\bar{\mu}, \bar{\nu})$, and r is that integer such that $\bar{\nu}_j = \nu_j$ for $j \leq r$ and $\bar{\nu}_j = \nu_j - 1$ for $r < j \leq q$.

If (μ, ν) is of type III, $(\bar{\mu}, \bar{\nu})$ is a (p, q)-shuffle. In case (ii), let $i = \mu_r$, $i-1 = \nu_s$ and define $\bar{\mu}_j = \mu_j$ for $j \neq r$, $\bar{\mu}_r = i-1$. In case (iii), let $i = \nu_s$, $i-1 = \mu_r$ and define $\bar{\mu}_j = \mu_j$ for $j \neq r$, $\bar{\mu}_r = i$.

Next, to each shuffle (μ, ν) and each i we associate a $(p+1, q)$-shuffle $(\tilde{\mu}, \tilde{\nu})$ and a second index t as follows. If s is the largest integer such that $\mu_j < i$ for $j < s$, then $\tilde{\mu}_j = \mu_j$ for $j < s$, $\tilde{\mu}_s = i$, and $\tilde{\mu}_j = \mu_{j-1} + 1$ for $j > s$. t is defined as the number of ν_j such that $i > \nu_j$.

DEFINITION 6.6: Let $F\colon K \times \Delta[q] \to L$ be a simplicial map. If (μ, ν) is a (p, q)-shuffle, define

$$h_{(\mu,\nu)}\colon K_p \to L_{p+q}$$

by

$$h_{(\mu,\nu)}(y) = F(s_{\nu_q} \dots s_{\nu_1} y, s_{\mu_p} \dots s_{\mu_1} \Delta_q).$$

If (μ, ν) is a $(p, q-1)$-shuffle and $i \in \{0, \dots, q\}$, define

$$h^i_{(\mu,\nu)}\colon K_p \to L_{p+q-1}$$

by

$$h^i_{(\mu,\nu)}(y) = F(s_{\nu_{q-1}} \dots s_{\nu_1} y, s_{\mu_p} \dots s_{\mu_1} \partial_i \Delta_q).$$

If $q = 1$ and we let $h_i = h_{(\mu,\nu)}$ where $\nu_1 = i$, then the proof of Proposition 6.2 states that h is a homotopy and that given a homotopy h we can construct a map F. A tedious verification proves the following generalization.

PROPOSITION 6.7: Let $F\colon K \times \Delta[q] \to L$ be a simplicial map. Then

(i) $\partial_i h_{(\mu,\nu)} = h^{i-r}_{(\bar{\mu},\bar{\nu})}$ if (μ, ν) is (p, q)-shuffle of type I and index r with respect to i;

(ii) $\partial_i h_{(\mu,\nu)} = h_{(\bar{\mu},\bar{\nu})} \partial_{i-r}$ if (μ, ν) is a (p, q)-shuffle of type II and index r with respect to i;

(iii) $\partial_i h_{(\mu,\nu)} = \partial_i h_{(\bar{\mu},\bar{\nu})}$ if (μ, ν) is a (p, q)-shuffle of type III with respect to i;

(iv) $s_i h_{(\mu,\nu)} = h_{(\tilde{\mu},\tilde{\nu})} s_{i-t}$ if (μ, ν) is a (p, q)-shuffle and t is the second index of (μ, ν) with respect to i.

Conversely, a set $\{h_{(\mu,\nu)}\}$ of functions $K_p \to L_{p+q}$ indexed on the (p, q)-shuffles and satisfying (ii)-(iv) determines a unique simplicial map $F: K \times \Delta[q] \to L$ by the first formula of Definition 6.6.

We will need one simple lemma to prove that L^K is a Kan complex if L is.

LEMMA 6.8: Let L be a Kan complex. Suppose given $r+1$ q-simplices $x_{i_0}, \ldots, x_{i_r}$ *of* L, where $r \leq q$ and $0 \leq i_0 < \cdots < i_r \leq q+1$ and suppose $\partial_{i_s} x_{i_t} = \partial_{i_t - 1} x_{i_s}$, $s < t$. Then there exists $x \in L_{q+1}$ such that $\partial_{i_s} x = x_{i_s}$, $0 \leq s \leq r$.

Proof: If $r = q$, the statement is true by the extension condition, and therefore the statement is true for $q = 0$. Assume $q > 0$ and the result holds for $q' < q$ and assume $r < q$ and the result holds for $r' > r$. Let $v \in \{0, \ldots, q+1\}$, $v \notin \{i_0, \ldots, i_r\}$ and let u be maximal such that $i_u < v$. Define $j_s = i_s$ if $i_s < v$, $j_{u+1} = v$, and $j_s = i_{s-1}$ if $u+1 < s \leq r+1$. We wish to find $x_v = x_{j_{u+1}}$ such that $x_{j_0}, \ldots, x_{j_{r+1}}$ satisfy the hypothesis, for then an application of the induction hypothesis on r gives the desired x. But the $(q-1)$-simplices $\partial_{v-1} x_{i_s}$, $s \leq u$, and $\partial_v x_{i_s}$, $u+1 \leq s \leq r$, satisfy the hypothesis, hence induction on q gives the desired x_v.

THEOREM 6.9: If K is a complex and L is a Kan complex, then L^K is a Kan complex.

Proof: Let $F_0, \ldots, F_{k-1}, F_{k+1}, \ldots, F_q$ be compatible $(q-1)$-simplices of L^K, and let $\{h^i_{(\mu,\nu)}\}$ be the functions indexed on the $(p, q-1)$-shuffles determined by the F_i, $i \neq k$. We will define functions $h_{(\mu,\nu)}$ indexed on the (p, q)-shuffles and satisfying (i)-(iv) of Proposition 6.7. It will follow that the corresponding map F has $\partial_i F = F_i$, $i \neq k$. We order all (p, q)-shuffles (q fixed) by letting an (r, q)-shuffle precede a (p, q)-shuffle if $r < p$ and by letting a (p, q)-shuffle (μ, ν) precede a (p, q)-shuffle (μ', ν') if $\mu_i = \mu_i'$ for $i < j$ and $\mu_j < \mu_j'$. We proceed by induction on the (p, q)-shuffles. The first such shuffle is the unique (o, q)-shuffle. If $y \in K_0$, choose $z \in L_q$ such that $\partial_i z = h^i_{(o, q-1)} y$, $i \neq k$. This is possible by the extension condition. Define $h_{(o, q)}\, y = z$. Now suppose that $h_{(\mu', \nu')}$ is defined for $(\mu', \nu') < (\mu, \nu)$.

Case 1: (μ, ν) is the first (p, q)-shuffle, so that $\mu_i = i-1$, $1 \leq i \leq p$, and $\nu_i = i+p-1$, $1 \leq i \leq q$. (μ, ν) is of type III with respect to p, and $(\mu, \nu) < (\bar{\mu}, \bar{\nu})$, since $\bar{\mu}_i = i-1$ for $i < p$ and $\bar{\mu}_p = p$. Thus if $y \in K_p$, $\partial_p h_{(\mu, \nu)} y$ has not yet been defined, but all other faces have been except $\partial_{p+k} h_{(\mu, \nu)}$ if $k > 1$. If y is non-degenerate, we may apply the lemma to define $h_{(\mu, \nu)}(y)$. If y is degenerate, we may use (iv) of Proposition 6.7 to define $h_{(\mu, \nu)}(y)$.

Case 2: $\mu_r = i-1$ and $\nu_s = i$, $r \in \{1, \ldots, p\}$, $s \in \{1, \ldots, q\}$. $(\mu, \nu) < (\bar{\mu}, \bar{\nu})$, where $(\bar{\mu}, \bar{\nu})$ is the associated (p, q)-shuffle relative to i, hence $\partial_i h_{(\mu, \nu)}$ has not been defined and we may proceed as in case 1.

Case 3: (μ, ν) is the last (p, q)-shuffle, so that $\mu_i = i + q - 1$, $1 \leq i \leq p$, and $\nu_i = i-1$, $1 \leq i \leq q$. Suppose also that $k < q$. Now $\partial_k h_{(\mu, \nu)} = h^k_{(\bar{\mu}, \bar{\nu})}$, where $(\bar{\mu}, \bar{\nu})$ is associated with (μ, ν) relative to k, and $h^k_{(\bar{\mu}, \bar{\nu})}$ has not been defined. Thus we may proceed as in the previous cases.

Case 4: Cases 1, 2, 3 complete the proof except when $k = q$. In this case, we simply reverse the ordering of the (p, q)-shuffles, p fixed, and proceed exactly as above starting with case 3, continuing with case 2 where now $\mu_r = i$ and $\nu_s = i-1$ for some i, r, and s, and finishing with case 1 where now $\partial_{p+q} h_{(\mu,\nu)}$ will not have been defined. Similarly we can prove

THEOREM 6.10: *If L' is a sub Kan complex of a Kan complex L, then $(L, L')^{(K, K')}$ is a Kan complex.*

As we noticed before, these results imply

COROLLARY 6.11: *Homotopy is an equivalence relation on maps into Kan complexes or Kan pairs.*

We remark that L^K is a functor, covariant in L and contravariant in K. The following two properties of this functor will be needed later.

LEMMA 6.12: *Let K, L, and M be complexes. Then we have a natural transformation of functors μ: $L^K \times M^L \to M^K$ defined by*

$$\mu(f, g)(x, u) = g(f(x, u), u), \quad x \in K, \quad u \in \Delta[n], \quad f \in L^K, \quad g \in M^L.$$

In particular, K^K is a simplicial monoid which operates on L^K from the left and on K^L from the right.

LEMMA 6.13: *Let K, L, and M be complexes. Then there exists a natural equivalence of functors ψ: $M^{K\times L} \to (M^K)^L$.*

Proof: Given f: $K\times L\times\Delta[n] \to M$, we define

$$\psi(f)(y, v)(x, u) = f(x, T(u)(y, v)),$$

where

$$x \in K_p, \quad y \in L_q, \quad u \in \Delta[q]_p, \quad v \in \Delta[n]_q$$

and where

T is the contravariant functor on Δ^* which defines $L\times\Delta[n]$.

Thus we are regarding u as an element of $\mathrm{Hom}(\Delta_p, \Delta_q)$ in the category Δ^* so that $T(u)$: $L_q\times\Delta[n]_q \to L_p\times\Delta[n]_p$. Next, given

$g\colon L \times \Delta[n] \to M^K$, we define

$$\phi(g)(x, y, v) = g(y, v)(x, \Delta_q), \quad x \in K_q, \quad y \in L_q, \quad v \in \Delta[n]_q$$

It is easy to see that ψ and ϕ are simplicial maps and are inverse isomorphisms.

In particular, $\psi|_{(M^{K\times L})_o}$ gives an isomorphism of sets $\psi_0\colon \operatorname{Hom}(K \times L, M) \to \operatorname{Hom}(L, M^K)$. Taking $L = M$, we find that $\psi_o(f)$, where $f\colon K \times L \to L$ is the projection, is an injection $L \to L^K$. Explicitly,

$$\psi_o(f)(y)(x, u) = \bar{y}(u), \quad y \in L_n, \quad x \in K, \quad u \in \Delta[n] .$$

BIBLIOGRAPHICAL NOTES ON CHAPTER I

Simplicial sets were first defined by Eilenberg and Zilber [15], who called them complete semi-simplicial complexes. The categorical definition of simplicial objects is due to Kan [33], and is also used by Cartan [6]. Kan [25] first recognized the possibility of defining the absolute homotopy groups of complexes satisfying the extension condition using only the simplicial structure. Our definition follows Kan [30]. A slightly different formulation is given in Moore [52]. The proof that $\pi_n(K, \phi)$ is Abelian, $n \geq 2$, is due to Moore [52]. The treatment of homotopy between maps and the proof that L^K is a Kan complex if L is, are both essentially as given in Moore [52]. Alternative proofs of the latter fact may be found in Cartan [6], Gugenheim [18], and MacLane [41]. Note that this theorem is the simplicial analog of Milnor's result [50] that if X has the homotopy type of a CW-complex and Y is compact, then X^Y has the homotopy type of a CW-complex. In the literature, L^K is often denoted by Hom (K, L), emphasizing the analogy with category theory rather than that with topology.

CHAPTER II

FIBRATIONS, POSTNIKOV SYSTEMS, AND MINIMAL COMPLEXES

§7. Kan fibrations

DEFINITION 7.1: Let $p\colon E \to B$ be a simplicial map. p is said to be a Kan fibration if for every collection of $n+1$ n-simplices $x_0, \dots, x_{k-1}, x_{k+1}, \dots, x_{n+1}$ of E which satisfy the compatibility condition $\partial_i x_j = \partial_{j-1} x_i$, $i < j$, $i \neq k$, $j \neq k$, and for every $(n+1)$-simplex y of B such that $\partial_i y = p(x_i)$, $i \neq k$, there exists an $(n+1)$-simplex x of E such that $\partial_i x = x_i$, $i \neq k$, and $p(x) = y$. E is called the total complex, B the base complex, and (E, p, B) is called a fibre space. If ϕ denotes the complex generated by a vertex of B, $F = p^{-1}(\phi)$ is called the fibre over ϕ. If ψ is the complex generated by a vertex of F, then the sequence

$$(F, \psi) \xrightarrow{i} (E, \psi) \xrightarrow{p} (B, \phi)$$

is called a fibre sequence.

REMARK 7.2: Let E be a complex, ϕ the complex generated by a point. Then the unique simplicial map $E \to \phi$ is a Kan fibration if and only if E is a Kan complex.

PROPOSITION 7.3: Let $p\colon E \to B$ be a Kan fibration. Then F is a Kan complex.

Proof: Suppose $x_0, \dots, x_{k-1}, x_{k+1}, \dots, x_{n+1}$ are compatible

n-simplices of F. Using our convention that ϕ denotes ambiguously any simplex of $\phi \subset B$, $p(x_i) = \phi$, and, since p is a Kan fibration, there exists $x \in E_{n+1}$ such that $\partial_i x = x_i$, $i \neq k$, and $p(x) = \phi$. But then $x \in F_{n+1}$, since by definition $F = p^{-1}(\phi)$.

LEMMA 7.4: Let $p\colon E \to B$ be a Kan fibration and assume p is onto. Suppose given $r+1$ q-simplices $x_{i_0}, \ldots, x_{i_r}$ of E, where $r \leq q$ and $0 \leq i_0 < \cdots < i_r \leq q+1$ and suppose

$$\partial_{i_s} x_{i_t} = \partial_{i_t - 1} x_{i_s}, \qquad s < t .$$

Assume further that $y \in B_{q+1}$ satisfies $\partial_{i_s} y = p(x_{i_s})$, $0 \leq s \leq r$. Then there exists $x \in E_{q+1}$ such that $\partial_{i_s} x = x_{i_s}$, $0 \leq s \leq r$, and $p(y) = x$.

The proof of the lemma is identical with that of Lemma 6.8, of which it is a generalization.

PROPOSITION 7.5: Let $p\colon E \to B$ be a Kan fibration.

(i) If E is a Kan complex and p is onto, then B is a Kan complex.

(ii) If B is a Kan complex, then E is a Kan complex.

Proof: (i) Suppose E is a Kan complex, and let

$$y_0, \ldots, y_{k-1}, y_{k+1}, \ldots, y_{q+1}$$

be q-simplices of B such that

$$\partial_i y_j = \partial_{j-1} y_i, \quad i < j, \quad i \neq k, \quad j \neq k .$$

Using the lemma, choose $x_0 \in E_q$ such that $p(x_0) = y_0$, choose $x_1 \in E_q$ such that $p(x_1) = y_1$ and $\partial_0 x_1 = \partial_0 x_0$, and so forth, until $x_0, \ldots, x_{k-1}, x_{k+1}, \ldots, x_{q+1}$ have been chosen satisfying $p(x_i) = y_i$, $i \neq k$, and $\partial_i x_j = \partial_{j-1} x_i$, $i < j$, $i \neq k$, $j \neq k$. Then we choose $x \in E_{q+1}$ such that $\partial_i x = x_i$, $i \neq k$, and we let $y = p(x)$. Obviously $\partial_i y = y_i$, $i \neq k$.

(ii) Suppose B is a Kan complex, and let

$$x_0, \ldots, x_{k-1}, x_{k+1}, \ldots, x_{q+1}$$

be q-simplices of E such that $\partial_i x_j = \partial_{j-1} x_i$, $i < j$, $i \neq k$, $j \neq k$. Choose $y \in B_{q+1}$ such that $\partial_i y = p(x_i)$, $i \neq k$. Then we may choose $x \in E_{q+1}$ such that $p(x) = y$ and $\partial_i x = x_i$, $i \neq k$.

Now let $(F, \psi) \xrightarrow{i} (E, \psi) \xrightarrow{p} (B, \phi)$ be a fibre sequence of Kan pairs. Regarding p as a map $(E, F, \psi) \to (B, \phi, \phi)$, p induces $p_\#: \pi_n(E, F, \psi) \to \pi_n(B, \phi)$. Let $y \in \tilde{B}_n$, $n \geq 1$. Since p is a Kan fibration, there exists $x \in E_n$ such that $\partial_i x = \psi$, $1 \leq i \leq n$, and $p(x) = y$. $\partial_0 x \in \tilde{F}_{n-1}$, since $p(\partial_0 x) = \partial_0 y = \phi$. Define q: $\pi_n(B, \phi) \to \pi_n(E, F, \psi)$ and $\partial_\#: \pi_n(B, \phi) \to \pi_{n-1}(F, \psi)$ by $q[y] = [x]$ and $\partial_\#[y] = [\partial_0 x]$. It is easily verified that q and $\partial_\#$ are well defined and are homomorphisms if $n \geq 2$. Further $p_\# q = 1$, $qp_\# = 1$, and the following diagram is commutative:

$$\begin{array}{ccccccccc}
\cdots \to & \pi_{n+1}(E, F, \psi) & \xrightarrow{\partial} & \pi_n(F, \psi) & \xrightarrow{i} & \pi_n(E, \psi) & \xrightarrow{j} & \pi_n(E, F, \psi) & \to \cdots \\
& \downarrow p_\# & & \| & & \| & & \downarrow p_\# & \\
\cdots \to & \pi_{n+1}(B, \phi) & \xrightarrow{\partial_\#} & \pi_n(F, \psi) & \xrightarrow{i} & \pi_n(E, \psi) & \xrightarrow{p_*} & \pi_n(B, \phi) & \to \cdots
\end{array}$$

Thus we have proven

THEOREM 7.6: Let $(F, \psi) \to (E, \psi) \to (B, \phi)$ be a fibre sequence of Kan pairs. Then $p_\#: \pi_n(E, F, \psi) \to \pi_n(B, \phi)$ is an isomorphism and the following is an exact sequence:

$$\cdots \to \pi_{n+1}(B, \phi) \xrightarrow{\partial_\#} \pi_n(F, \psi) \xrightarrow{i} \pi_n(E, \psi) \xrightarrow{p_*} \pi_n(B, \phi) \to \cdots .$$

Next we define maps and homotopies of maps of fibre spaces.

DEFINITION 7.7: Let (E, p, B) and (E', p', B') be fibre spaces. A map $(\tilde{f}, f)$: $(E, p, B) \to (E', p', B')$ is a commutative diagram

$$\begin{array}{ccc} E & \xrightarrow{\tilde{f}} & E' \\ {\scriptstyle p}\downarrow & & \downarrow{\scriptstyle p'} \\ B & \xrightarrow{f} & B' \end{array}$$

of simplicial complexes. A homotopy $(\tilde{F}, F)$: $(\tilde{f}, f) \simeq (\tilde{g}, g)$ is a commutative diagram

$$\begin{array}{ccc} E \times I & \xrightarrow{\tilde{F}} & E' \\ \downarrow{\scriptstyle p\times 1} & & \downarrow{\scriptstyle p'} \\ B \times I & \xrightarrow{F} & B' \end{array}$$

of simplicial complexes, where F: $f \simeq g$ and $\tilde{F}$: $\tilde{f} \simeq \tilde{g}$. (E, p, B) is said to be the same homotopy type as (E', p', B') if there exist $(\tilde{f}, f)$: $(E, p, B) \to (E', p', B')$ and $(\tilde{g}, g)$: $(E', p', B') \to (E, p, B)$ such that $(\tilde{f}, f)(\tilde{g}, g) \cong 1$ and $(\tilde{g}, g)(\tilde{f}, f) \cong 1$. A sub-fibre space (E, p, B) of (E', p', B') is said to be a deformation retract of (E', p', B') if the identity of (E', p', B') is homotopic relative to (E, p, B) to a map of (E', p', B') onto (E, p, B) which extends the inclusion map $(E, p, B) \to (E', p', B')$.

Using Lemma 7.4 and the machinery developed to prove Theorem 6.9, we can prove the following generalization of that theorem.

THEOREM 7.8: If p: $E \to B$ is a Kan fibration, then so is p^K: $E^K \to B^K$, where K is any complex. If F is the fibre over $\phi \subset B$, then F^K is the fibre over $\phi^K \cong \phi$.

COROLLARY 7.9: Let (E, p, B) and (E_1, p_1, B_1) be fibre spaces and suppose that B_1 is a Kan complex. Then homotopy is an equivalence relation on maps $(E, p, B) \to (E_1, p_1, B_1)$.

Proof: Given $(\tilde{f}, f)$, define $\tilde{F}(x, z) = \tilde{f}(x)$ and $F(y, z) = f(y)$, $x \in E_n$, $y \in B_n$, $z \in I_n$.

Then $(\tilde{F}, F)$: $(\tilde{f}, f) \simeq (\tilde{f}, f)$. Now suppose $(\tilde{F}, F)$: $(\tilde{f}, f) \simeq (\tilde{g}, g)$ and $(\tilde{G}, G)$: $(\tilde{f}, f) \simeq (\tilde{h}, h)$. We must prove $(\tilde{g}, g) \simeq (\tilde{h}, h)$. ${B_1}^B$ is a Kan complex, hence there exists $U \in ({B_1}^B)_2$ such that $\partial_1 U = G$ and $\partial_2 U = F$. Let $H = \partial_0 U$. Then H: $g \simeq h$. Since

$$p_1^E: E_1^E \to B_1^E$$

is a Kan fibration, there exists $V \in ({E_1}^E)_2$ such that $\partial_1 V = \tilde{G}$, $\partial_2 V = \tilde{F}$, and $p_1^E(V) = U \circ (p \times 1)$. Let $\tilde{H} = \partial_0 V$. Then $\tilde{H}$: $\tilde{g} \simeq \tilde{h}$ and $p_1 \circ \tilde{H} = p_1^E(\tilde{H}) = H \circ (p \times 1)$ as desired.

To prove the next corollary, we must first rectify a small omission.

DEFINITION 7.10: Two vertices x and y of a complex B are in the same path component of B, written $x \sim y$, if there exist 1-simplices $z_1, \ldots, z_n$ such that x is a face of z_1, y is a face of z_n, and z_i has a face in common with z_{i+1}, $1 \leq i < n$. $\sim$ is an equivalence relation on vertices of B, and we define

$$\pi_0(B) = B_0/(\sim).$$

If B is a Kan complex, then obviously $\pi_0(B) = \pi_0(B, \phi)$, where ϕ is any vertex of B.

COROLLARY 7.11: Let p: $E \to B$ be a Kan fibration and let F_ϕ and F_ψ be the fibres over two vertices ϕ and ψ of B. Then if ϕ and ψ are in the same path component of B, F_ϕ and F_ψ are of the same homotopy type.

Proof: p^{F_ϕ}: $E^{F_\phi} \to B^{F_\phi}$ is a Kan fibration. As shown in Remarks 6.10, there is an injection $B \to B^{F_\phi}$, and if we let $K = (p^{F_\phi})^{-1}(B)$ and $q = p^{F_\phi} | K$, then q: $K \to B$ is a Kan fibration. An element $x \in K_n$ is a map $F_\phi \times \Delta[n] \to E$ for which there exists a commutative diagram

$$\begin{array}{ccc} F_\phi \times \Delta[n] & \xrightarrow{x} & E \\ \downarrow & & \downarrow p \\ \Delta[n] & \longrightarrow & B \end{array}$$

In particular, an element of K_0 is a map from F_ϕ into another fibre. Now we may assume that there exists $z \in B_1$ such that $\partial_1 z = \phi$ and $\partial_0 z = \psi$. If $x \in K_0$ is the inclusion $F_\phi \to E$, then there exists $y \in K_1$ such that $q(y) = z$ and $\partial_1 y = x$. It follows that $\partial_0 y$ is a map $F_\phi \to F_\psi$. Similarly, let $K' = (p^{F_\psi})^{-1}(B)$, $q' = p^{F_\psi} | K'$: $K' \to B$. If x': $F_\psi \to E$ is the inclusion there exists $y' \in K_1'$ such that $q(y') = z$, $\partial_0 y' = x'$, and $\partial_1 y'$ is a map $F_\psi \to F_\phi$. Since $q: K \to B$ is a Kan fibration, there exists $w \in K_2$ such that $\partial_0 w = y$, $\partial_1 w = y' \circ (\partial_0 y \times 1)$, and $q(w) = s_0 z$. It follows that $q(\partial_2 w) = \phi$, $\partial_0 \partial_2 w = x$, and $\partial_1 \partial_2 w = \partial_1 y' \circ \partial_0 y$. Since $\partial_2 w (F_\phi \times I) \subset F_\phi$, $\partial_2 w$: $\partial_1 y' \circ \partial_0 y \simeq 1$ in F_ϕ. Similarly, we may prove that $\partial_0 y \circ \partial_1 y' \simeq 1$ in F_ψ, and therefore F_ϕ and F_ψ are of the same homotopy type.

COROLLARY 7.12: (Covering Homotopy Property). Let p: $E \to B$ be a Kan fibration and let K be any complex. Let $\tilde{f}$: $K \to E$ and $f = p \circ \tilde{f}$: $K \to B$. Suppose F: $K \times I \to B$ satisfies $\partial_1 F = f$. Then there exists $\tilde{F}$: $K \times I \to E$ such that $p \circ \tilde{F} = F$ and $\partial_1 \tilde{F} = \tilde{f}$.

A proof similar to that of Theorem 6.9 also gives

THEOREM 7.13: Let i: $K \to L$ be an inclusion of complexes and let E be a Kan complex. Then E^i: $E^L \to E^K$ is a Kan fibration.

COROLLARY 7.14: (Homotopy Extension Theorem). Let K be a subcomplex of the complex L and let E be a Kan complex. Let $\tilde{f}$: $L \to E$ and let $f = \tilde{f}|_K$: $K \to E$. Suppose F: $K \times I \to E$ satisfies $\partial_1 F = f$. Then there exists $\tilde{F}$: $L \times I \to E$ such that $\tilde{F}|K \times I = F$ and $\partial_1 \tilde{F} = \tilde{f}$.

Theorems 7.8 and 7.13 are both special cases of a more general result, which we now state.

DEFINITION 7.15: A commutative diagram of complexes and maps

$$\begin{array}{ccc} A & \xrightarrow{e} & B \\ {\scriptstyle f}\downarrow & & \downarrow{\scriptstyle g} \\ C & \xrightarrow{h} & D \end{array}$$

is said to be a fibre square if for any $q+1$ compatible q-simplices $x_0, \ldots, x_{k-1}, x_{k+1}, \ldots, x_{q+1}$ of A and for any $y \in B_{q+1}$ and $z \in C_{q+1}$ such that $\partial_i y = e(x_i)$ and $\partial_i z = f(x_i)$, $i \neq k$, and $g(y) = h(z)$, there exists $x \in A_{q+1}$ such that $e(x) = y$, $f(x) = z$, *and* $\partial_i x = x_i$, $i \neq k$.

THEOREM 7.16: Let $p\colon E \to B$ be a Kan fibration and let $i\colon K \to L$ be an inclusion of complexes. Then

$$\begin{array}{ccc} E^L & \xrightarrow{p^L} & B^L \\ {\scriptstyle E^i}\downarrow & & \downarrow{\scriptstyle B^i} \\ E^K & \xrightarrow{p^K} & B^K \end{array}$$

is a fibre square.

COROLLARY 7.17: (Covering homotopy extension theorem). Suppose given $(\tilde{f}, f) : (L, g, L') \to (E, p, B)$, where g is any map of complexes and p is a Kan fibration. Let $K' \subset L'$ and

$$K = g^{-1}(K') \subset L$$

(here K' may be empty). Further, let $G\colon K \times I \to E$ and $F\colon L' \times I \to B$ be given satisfying $\partial_1 F = f$, $\partial_1 G = \tilde{f}|K$, and $p \circ G = F \circ (g|K)$. Then there exists $\tilde{F}\colon L \times I \to E$ such that $\tilde{F}|K \times I = G$, $p \circ \tilde{F} = F \circ (g \times 1)$, and $\partial_1 \tilde{F} = \tilde{f}$.

§ 8. Postnikov systems

In this section we define the natural Postnikov systems

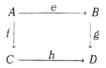

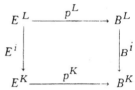

associated with Kan complexes and fibre spaces. The definition of the k-invariants will be given later, after the concept of twisted Cartesian product has been developed.

DEFINITIONS 8.1: Let $\Delta[q]^n$ denote the n-skeleton of $\Delta[q]$, that is, the subcomplex generated by all simplices of dimension $\leq n$. Define an equivalence relation $\overset{n}{\sim}$ on q-simplices of a complex K by $x \overset{n}{\sim} y$ if $\bar{x}|\Delta[q]^n = \bar{y}|\Delta[q]^n$ (where $\bar{x}(\Delta_q) = x$, $\bar{y}(\Delta_q) = y$). Noting that $x \overset{n}{\sim} y$ implies $\partial_i x \overset{n}{\sim} \partial_i y$ and $s_i x \overset{n}{\sim} s_i y$, we can define a complex $K^{(n)}$ by $K_q^{(n)} = K_q/(\overset{n}{\sim})$, where the face and degeneracy operators are induced from those of K. Note that $K_q^{(n)} = K_q$ if $n \geq q$, and define $K^{(\infty)} = K$. Let p_m^n denote the natural map of $K^{(n)}$ onto $K^{(m)}$, $m \leq n \leq \infty$. We will usually write p for p_m^n.

PROPOSITION 8.2: Let K be a Kan complex. Then

(i) $K^{(n)}$ is a Kan complex;

(ii) $p_m^n: K^{(n)} \to K^{(m)}$ is a Kan fibration.

Proof: (i) follows from Proposition 7.5 and the case $n = \infty$ of (ii). Let $x_0, \ldots, x_{k-1}, x_{k+1}, \ldots, x_{q+1}$ be compatible q-simplices of $K^{(n)}$ and suppose $y \in K_{q+1}^{(m)}$ satisfies $\partial_i y = p(x_i)$, $i \neq k$. If $q \leq m$, $K_q^{(m)} = K_q$ so that if $z \in K_{q+1}^{(n)}$ represents y, $\partial_i z = x_i$, $i \neq k$. Then if we let $x = z$, $\partial_i x = x_i$, $i \neq k$, and $p(x) = y$. Thus assume $q > m$. Suppose first that $n = \infty$. Then there exists $x \in K_{q+1}$ such that $\partial_i x = x_i$, $i \neq k$. Since any face of x of dimension $\leq m$ is also a face of some x_i, $x \overset{m}{\sim} z$ if z represents y and therefore $p(x) = y$. Now if $n < \infty$, we know that $K^{(n)}$ is a Kan complex, and the same argument applies.

DEFINITION 8.3: Let (K, ϕ) be a Kan pair. Define $E_n(K, \phi)$ to be a fibre over ϕ of p: $K \to K^{(n-1)}$. Thus $E_n(K, \phi)$ is the subcomplex of K consisting of those simplices of K all of whose faces of dimension less than n are at ϕ. $E_n(K, \phi)$ is called the nth Eilenberg subcomplex of K based at ϕ.

THEOREM 8.4: Let (K, ϕ) be a Kan pair. Then

(i) $p_*\colon \pi_q(K^{(n)}, \phi) \xrightarrow{\cong} \pi_q(K^{(m)}, \phi)$ for $q \leq m$;

(ii) $\pi_q(K^{(m)}, \phi) = 0$ for $q > m$;

(iii) $i\colon \pi_q(E_{m+1}(K^{(n)}, \phi), \phi) \xrightarrow{\cong} \pi_q(K^{(n)}, \phi)$ for $q > m$;

(iv) $\pi_q(E_{m+1}(K^{(n)}, \phi), \phi) = 0$ for $q \leq m$ and for $q > n$.

Proof: $E_{m+1}(K^{(n)}, \phi)_q = \phi$ if $q \leq m$. This proves part of (iv) and the homotopy exact sequence of the fibre space $(K^{(n)}, p, K^{(m)})$ then gives (i). Suppose $x \in a \in \pi_q(K^{(m)}, \phi)$, $q > m$. By definition, $\partial_i x = \phi$ for all i and, again by definition, this implies that ϕ is a representative for x. Therefore $a = 0$, and (ii) is proven. Using the homotopy exact sequence, (ii) implies (iii). The last part of (iv) now follows.

The special cases $n = \infty$ and $n = m+1$ are of particular importance.

DEFINITION 8.5: If K is a Kan complex, let X^n denote $(K^{(n+1)}, p, K^{(n)})$. The sequence of fibre spaces $X = (X^0, \ldots, X^n, \ldots)$ is called the natural Postnikov system of K. We let $F^{(n+1)}(X)$ denote the fibre $E_{n+1}(K^{(n+1)}, \phi)$ of the fibre space X^n.

DEFINITION 8.6: If (K, ϕ) is a connected Kan pair such that for some $n > 0$, $\pi_n(K, \phi) = \pi$ and $\pi_q(K, \phi) = 0$ if $q \neq n$, then K is said to be an Eilenberg-MacLane complex of type (π, n).

The fundamental property of the natural Postnikov system is given by the following corollary of Theorem 8.4.

COROLLARY 8.7: Let (K, ϕ) be a Kan pair, X its natural Postnikov system. Then $F^{(n)}(X)$ is an Eilenberg-MacLane complex of type $(\pi_n(K, \phi), n)$, $n \geq 1$.

Now we generalize the constructions above to fibre sequences. For the remainder of this section, $(F, \psi) \xrightarrow{i} (E, \psi) \xrightarrow{p} (B, \phi)$ will be a fibre sequence of connected Kan pairs. We will define the natural Postnikov system of the fibre space (E, p, B).

DEFINITIONS 8.8: Define an equivalence relation $\overset{n}{\sim}$ on q-simplices of E by $x \overset{n}{\sim} y$ if $\bar{x}|\Delta[q]^n = \bar{y}|\Delta[q]^n$ and $p(x) = p(y)$. Define a complex $E^{(n)}$ by $E^{(n)}_q = E_q/(\overset{n}{\sim})$, where the face and degeneracy operators are induced from those of E, and let $E^{(\infty)} = E$. Let p^n_m denote the natural map of $E^{(n)}$ onto $E^{(m)}$, $m \leq n \leq \infty$, and write p for p^n_m. Note that if $E_0 = \psi$, then $E^{(0)} \cong B$, since $x \overset{0}{\sim} y$ if $p(x) = p(y)$.

PROPOSITION 8.8: Under the above hypotheses

(i) $E^{(n)}$ is a Kan complex;

(ii) $p\colon E^{(n)} \to B$ is a Kan fibration;

(iii) $p^n_m\colon E^{(n)} \to E^{(m)}$ is a Kan fibration.

THEOREM 8.9: Under the above hypotheses

(i) the fibre over ϕ of $p\colon E^{(n)} \to B$ is $F^{(n)}$, the nth total space in the natural Postnikov system for the fibre F;

(ii) the fibre over ψ of $p^n_m\colon E^{(n)} \to E^{(m)}$ is $E_{m+1}(F^{(n)}, \psi)$;

(iii) $p_*\colon \pi_q(E^{(n)}, \psi) \xrightarrow{\cong} \pi_q(B, \phi)$ for $q > n+1$;

(iv) $p_*\colon \pi_q(E^{(n)}, \psi) \xrightarrow{\cong} \pi_q(E^{(m)}, \psi)$ for $q \leq m$ and for $q > n+1$.

DEFINITION 8.10: Let $\mathcal{E}^n$ denote $(E^{(n+1)}, p, E^{(n)})$. The sequence of fibre spaces $\mathcal{E} = (\mathcal{E}^0, \ldots, \mathcal{E}^n, \ldots)$ is called the natural Postnikov system of (E, p, B). Let $F^{(n+1)}(\mathcal{E})$ denote the fibre of $\mathcal{E}^n$.

COROLLARY 8.11: $F^{(n)}(\mathcal{E}) = F^{(n)}(X)$, where X is the natural Postnikov system of the fibre F, and therefore $F^{(n)}(\mathcal{E})$ is an Eilenberg-MacLane complex of type $(\pi_n(F,\psi), n)$, $n \geq 1$.

REMARK 8.12: The important part of the homotopy exact sequence of $\mathcal{E}^n$ is the following:

$$0 \to \pi_{n+2}(E^{(n+1)}, \psi) \to \pi_{n+2}(E^{(n)}, \psi) \to \pi_{n+1}(F^{(n+1)}(\mathcal{E}), \psi)$$
$$\to \pi_{n+1}(E^{(n+1)}, \psi) \to \pi_{n+1}(E^{(n)}, \psi) \to 0.$$

This is essentially just the exact sequence

$$0 \to p_*(\pi_{n+2}(E,\psi)) \to \pi_{n+2}(B,\phi) \to \pi_{n+1}(F,\psi) \to \pi_{n+1}(E,\psi)$$
$$\to p_*(\pi_{n+1}(E,\psi)) \to 0.$$

§9. Minimal complexes

DEFINITION 9.1: A Kan complex K is said to be minimal if $\partial_i x = \partial_i y$, $i \neq k$, implies $\partial_k x = \partial_k y$.

LEMMA 9.2: A Kan complex K is minimal if and only if $x \sim y$ implies $x = y$.

Proof: (i) Suppose K is minimal and $x \sim y$, $x, y \in K_q$. Then there exists $z \in K_{q+1}$ such that $\partial_i z = s_{q-1}\partial_i x$, $i < q$, $\partial_q z = x$ and $\partial_{q+1} z = y$. Now $\partial_i s_q x = s_{q-1}\partial_i x$, $i < q$, and $\partial_q s_q x = x$. By assumption, this implies $\partial_{q+1} z = \partial_{q+1} s_q x$, that is, $y = x$.

(ii) Suppose $x \sim y$ implies $x = y$. Let $\partial_i x = \partial_i y$, $i \neq k$, $x, y \in K_{q+1}$. We must prove $\partial_k x = \partial_k y$, and it suffices to prove $\partial_k x \sim \partial_k y$, By the extension condition, if $k \leq q$ there exists $z \in K_{q+2}$ such that $\partial_i z = s_q \partial_i x$, $i \leq q$, $i \neq k$, $\partial_{q+1} z = x$, and $\partial_{q+2} z = y$. Then $\partial_k z : \partial_k x \sim \partial_k y$. If $k = q+1$, there exists $z \in K_{q+2}$ such that $\partial_i z = s_{q-1}\partial_i x$, $i < q$, $\partial_q z = x$, and $\partial_{q+1} z = y$. Then $\partial_{q+2} z\colon \partial_{q+1} x \sim \partial_{q+1} y$.

LEMMA 9.3: Let K be a complex. Suppose $x, y \in K_q$ satisfy $\partial_i x = \partial_i y$, all i, and $x = s_m z$, $y = s_n w$. Then $x = y$.

Proof: If $m = n$, $z = \partial_m x = \partial_m y = w$, and $x = y$. Assume $m < n$. $z = \partial_m x = \partial_m s_n w = s_{n-1} \partial_m w$, hence

$$x = s_m s_{n-1} \partial_m w = s_n s_m \partial_m w \, .$$

Thus $s_m \partial_m w = \partial_n x = \partial_n y = w$, and, finally, $x = s_n w = y$.

Now if K is a Kan complex, we define a minimal subcomplex of K as follows. Choose a vertex in each component of K. These vertices are the elements of M_0. Suppose $M_{q'}$ has been defined, $0 \le q' < q$. Consider the homotopy classes of q-simplices all of whose faces are in M_{q-1}, and choose one q-simplex in each such class, choosing a degenerate representative whenever possible. These q-simplices are the elements of M_q. By Lemma 9.3, M is a subcomplex of K. We will prove that M is a retract of K, and since any retract of a Kan complex is clearly a Kan complex, it will follow from Lemma 9.2 that M is a minimal complex.

DEFINITION 9.4: A deformation retract K of a complex L is said to be a strong deformation retract of L if some homotopy $F\colon L \times I \to L$ defining K as a deformation retract satisfies $F(x, y) = x$, $x \in K_q$, all q.

THEOREM 9.5: M is a strong deformation retract of K.

Proof: We will define h_i satisfying the conditions of Definition 5.1, with $\partial_0 h_0(x) = x$, and $\partial_{q+1} h_q(x) \in M_q$ for $x \in K_q$ and with $h_i(x) = s_i x$ for $x \in M_q$.

(i) If $x \in K_0$, let $h_0(x)$ be a 1-simplex satisfying

$\partial_0 h_0(x) = x$ and $\partial_1 h_0(x) \in M_0$, and let $h_0(x) = s_0 x$ if $x \in M_0$.

(ii) Assume h_i has been defined on $K_{q'}$, $0 \leq q' < q$, and let $x \in K_q$. If x is degenerate, say $x = s_j y$, define

$$h_i(x) = s_j h_{i-1}(y) \text{ if } j \leq i-1 \text{ and } h_i(x) = s_{j+1} h_i(y) \text{ if } j > i-1,$$

checking that if $x = s_k z$ is also true, the two resulting definitions agree. If $x \in M_q$, let $h_i(x) = s_i x$. Otherwise, let $h_0(x)$ be a $(q+1)$-simplex satisfying $\partial_0 h_0(x) = x$ and $\partial_i h_0(x) = h_0(\partial_{i-1} x)$ if $i > 1$. There exists such a simplex by the extension condition.

(iii) Assume h_i has been defined on K_q, $0 \leq i < j < q$, and let $x \in K_q$ be non-degenerate, $x \notin M_q$. By the extension condition, we can choose $h_j(x)$ satisfying $\partial_i h_j(x) = h_{j-1}(\partial_i x)$ for $i < j$, $\partial_j h_j(x) = \partial_j h_{j-1}(x)$, and $\partial_i h_j(x) = h_j(\partial_{i-1} x)$ for $i > j+1$.

(iv) It remains to define $h_q(x)$, where $x \in K_q$ is non-degenerate and $x \notin M_q$. Choose $y \in K_{q+1}$ such that

$$\partial_i y = h_{q-1}(\partial_i x), \quad i < q,$$

and

$$\partial_q y = \partial_q h_{q-1}(x).$$

Now if $i < q$, $\partial_i \partial_{q+1} y = \partial_q h_{q-1}(\partial_i x) \in M_{q-1}$, and

$$\partial_q \partial_{q+1} y = \partial_q h_{q-1}(\partial_q x) \in M_{q-1}.$$

Therefore there exists a unique $z \in M_q$ such that $z \sim \partial_{q+1} y$. By the extension condition, there exists $w \in K_{q+2}$ such that

$$\partial_i w = s_{q+1} \partial_i y, \quad 0 \leq i \leq q,$$

and

$$\partial_{q+1} w = s_{q+1} z.$$

Define $h_q(x) = \partial_{q+2} w$. It follows that $\partial_i h_q(x) = \partial_i y$, $0 \leq i \leq q$, and $\partial_{q+1} h_q(x) = z \in M_q$ as desired.

LEMMA 9.6: Let $f \simeq g\colon L \to M$, where M is a minimal complex. If f is an isomorphism, then so is g.

Proof: Let $h\colon L \to M$ be the given homotopy.

(i) Suppose $g(x) = g(y)$, $x, y \in L_q$. We must prove that $x = y$. Now $\partial_{q+1} h_q(x) = g(x) = g(y) = \partial_{q+1} h_q(y)$. If $q = 0$, then by the minimality of M $f(x) = \partial_0 h_0(x) = \partial_0 h_0(y) = f(y)$, hence $x = y$. Assume the result for $q' < q$, $q > 0$. Then $\partial_i x = \partial_i y$. If $i < q$,

$$\partial_i h_q(x) = h_{q-1}(\partial_i x) = h_{q-1}(\partial_i y) = \partial_i h_q(y) .$$

By the minimality of M, $\partial_q h_q(x) = \partial_q h_q(y)$. Now

$$\partial_i h_{q-1}(x) = \partial_i h_{q-1}(y), \quad i \neq q-1,$$

hence

$$\partial_{q-1} h_{q-1}(x) = \partial_{q-1} h_{q-1}(y) .$$

Iterating, we find $f(x) = \partial_0 h_0(x) = \partial_0 h_0(y) = f(y)$, and therefore $x = y$.

(ii) Let $y \in M_q$. We must find $x \in L_q$ such that $g(x) = y$. If $q = 0$, there exists $z \in M_1$ such that $\partial_1 z = y$. Let x satisfy $f(x) = \partial_0 z$. $\partial_0 h_0(x) = f(x) = \partial_0 z$, hence

$$g(x) = \partial_1 h_0(x) = \partial_1 z = y .$$

Assume the result for $q' < q$, $q > 0$. Let $x_i \in L_{q-1}$ be the unique element satisfying $g(x_i) = \partial_i y$. Choose $z_q \in M_{q+1}$ such that $\partial_{q+1} z_q = y$ and $\partial_i z_q = h_{q-1}(x_i)$, $i < q$. By induction on $q-j$, we may choose $z_j \in M_{q+1}$ such that $\partial_i z_j = h_{j-1}(x_i)$ if $i < j$, $\partial_{j+1} z_j = \partial_{j+1} z_{j+1}$, and $\partial_i z_j = h_j(x_{i-1})$ if $i > j+1$. Now let $x \in L_q$ satisfy $f(x) = \partial_0 z_0$. If $i > 0$,

$$f(\partial_i x) = \partial_i \partial_0 z_0 = \partial_0 \partial_{i+1} z_0 = \partial_0 h_0(x_i) = f(x_i) ,$$

and $f(\partial_0 x) = \partial_0 \partial_1 z_0 = \partial_0 \partial_1 z_1 = \partial_0 \partial_0 z_1 = \partial_0 h_0(x_0) = f(x_0)$.

Therefore $\partial_i x = x_i$ for all i. Now $\partial_i h_0(x) = \partial_i z_0$, $i \neq 1$, hence $\partial_1 h_0(x) = \partial_1 z_0$. Similarly, we find $\partial_i h_j(x) = \partial_i z_j$ for all i and j and, in particular, $g(x) = \partial_{q+1} h_q(x) = \partial_{q+1} z_q = y$, as desired.

PROPOSITION 9.7: Let $f\colon M \to M'$ be a homotopy equivalence, where M and M' are minimal complexes. Then f is an isomorphism.

Proof: Let $g\colon M' \to M$ satisfy $fg \simeq 1$ and $gf \simeq 1$. By the lemma, fg and gf are isomorphisms, hence so are g and f (although g need not be f^{-1}).

THEOREM 9.8: Any two minimal complexes which are deformation retracts of a Kan complex K are isomorphic.

Proof: Let M and M' be two such complexes with retractions $r\colon K \to M$ and $r'\colon K \to M'$ and let $i\colon M \subset K$ and $i'\colon M' \subset K$. Since $ir \simeq 1$ and $i'r' \simeq 1$ on K, $(r'i)(ri') \simeq r'i' = 1$ on M' and $(ri')(r'i) \simeq ri = 1$ on M, hence the result follows from the proposition.

CONVENTION 9.9: From now on, when we speak of a minimal subcomplex of a Kan complex K, we mean one which is a strong deformation retract of K.

We now discuss briefly a more general concept than that of minimal complex which applies to the relative case.

DEFINITIONS 9.10: A sub Kan complex M of a Kan complex K is said to be admissible if for every $x \in K_q$ such that $\partial_i x \in M_{q-1}$ for all i, there exists $y \in M_q$ such that $x \sim y$. If (K, L, ϕ) is a Kan triple, an admissible subcomplex M of K is said to be relatively admissible if $\phi \subset M$ and $M \cap L$ is an admissible subcomplex of L.

THEOREM 9.11: Let (K, L, ϕ) be a Kan triple and let M be a relatively admissible subcomplex of K. Then $(M, M \cap L, \phi)$ is a strong deformation retract of (K, L, ϕ).

The proof is essentially the same as that of Theorem 9.5.

REMARKS 9.12: A Kan triple (K, L, ϕ) is said to be n-connected if $\pi_i(K, L, \phi) = 0$, $1 \leq i \leq n$. The n-th relative Eilenberg subcomplex $E_n(K, L, \phi)$ of K is defined by $x \in E_n(K, L, \phi)_q$ if and only if $\overline{x}|\Delta[q]^0 = \overline{\phi}|\Delta[q]^0$ and $\text{Image}(\overline{x}|\Delta[q]^{n-1}) \subset L$. If (K, L, ϕ) is n-connected, then $E_{n+1}(K, L, \phi)$ is relatively admissible.

§10. Minimal fibrations

In this section, we outline the construction of minimal sub fibrations of Kan fibrations.

DEFINITIONS 10.1: A Kan fibration p: $E \to B$ is said to be minimal if $p(x) = p(y)$ and $\partial_i x = \partial_i y$, $i \neq k$, implies $\partial_k x = \partial_k y$. If p is minimal and B is a minimal complex, then (E, p, B) is said to be a minimal fibre space.

LEMMA 10.2: Let p: $E \to B$ be a Kan fibration. Then if p is minimal, each fibre of p is a minimal complex.

LEMMA 10.3: Let (E, p, B) be a fibre space. Then if E is a minimal complex and p is onto, (E, p, B) is a minimal fibre space.

Proof: Clearly p is a minimal fibration. We must prove that B is a minimal complex. Suppose $x, y \in B_q$ and z: $x \sim y$. Let $x = p(w)$. Since p is a Kan fibration, there exists $u \in E_{q+1}$ such that $\partial_i u = s_{q-1} \partial_i w$, $i < q$, $\partial_q u = w$, and $p(u) = z$. Now u: $\partial_q u \sim \partial_{q+1} u$, hence since E is minimal, $\partial_q u = \partial_{q+1} u$. $p(\partial_{q+1} u) = \partial_{q+1} z = y$ and $p(\partial_q u) = p(w) = x$, and therefore $x = y$.

DEFINITION 10.4: Let $p\colon E \to B$ be a map. If $x, y \in E_q$, we say that x is p-homotopic to y, written $x \overset{(p)}{\sim} y$, if $x \sim y$ and there exists $z \in E_{q+1}$ such that $z\colon x \sim y$ and $p(z) = s_q p(x)$.

LEMMA 10.5: If $p\colon E \to B$ is a Kan fibration, then $\overset{(p)}{\sim}$ is an equivalence relation on the q-simplices of E.

LEMMA 10.6: A Kan fibration $p\colon E \to B$ is minimal if and only if $x \overset{(p)}{\sim} y$ implies $x = y$.

Now if $p\colon E \to B$ is a Kan fibration, we can define a subcomplex E' of E by a procedure precisely parallel to the construction of a minimal subcomplex of a Kan complex, the only difference being that the relation of p-homotopy rather than that of homotopy is used.

LEMMA 10.7: $p\colon E' \to B$ is a Kan fibration.

It follows immediately from Lemma 10.6 that $p\colon E' \to B$ is a minimal fibration.

DEFINITION 10.8: A deformation retract (E', p, B') of a fibre space (E, p, B) is said to be a strong deformation retract if some homotopy $(\tilde{F}, F)$ defining (E', p, B') as a deformation retract satisfies

$$\tilde{F}(x, y) = x, \quad x \in E_q', \quad y \in I_q, \quad \text{all } q,$$

and

$$F(x, y) = x, \quad x \in B_q', \quad y \in I_q, \quad \text{all } q$$

THEOREM 10.9: (E', p, B) is a strong deformation retract of (E, p, B).

More generally, we can prove

THEOREM 10.10: Let $p\colon E \to B$ be a Kan fibration and let B' be a strong deformation retract of B. Then there exists $E' \subset p^{-1}(B')$ such that $p'\colon E' \to B'$ is a minimal fibration, and (E', p', B') is a strong deformation retract of (E, p, B).

COROLLARY 10.11: Let (E, p, B) be a fibre space of Kan complexes. Then there exists a minimal fibre space (E', p', B') which is a strong deformation retract of (E, p, B).

LEMMA 10.12: Suppose given a strong homotopy

$$(\tilde{f}, 1) \simeq (\tilde{g}, 1): (E', p', B) \to (E, p, B),$$

where p is a minimal fibration. Then if $\tilde{f}$ is an isomorphism, so is $\tilde{g}$.

PROPOSITION 10.13: Let $(\tilde{f}, 1): (E', p', B) \to (E, p, B)$ be a strong homotopy equivalence, where p and p' are minimal fibrations. Then $\tilde{f}$ is an isomorphism.

THEOREM 10.14: Any two minimal fibrations with base complex B which are strong deformation retracts of a given fibre space (E, p, B) are strongly isomorphic (in the sense that the isomorphism on B is the identity).

LEMMA 10.15: Let $(\tilde{f}, f) \simeq (\tilde{g}, g): (E', p', B') \to (E, p, B)$, where (E, p, B) is a minimal fibre space. Then if $(\tilde{f}, f)$ is an isomorphism, so is $(\tilde{g}, g)$.

PROPOSITION 10.16: Let $(\tilde{f}, f): (E', p', B') \to (E, p, B)$ be a homotopy equivalence, where (E', p', B') and (E, p, B) are minimal fibre spaces. Then $(\tilde{f}, f)$ is an isomorphism.

THEOREM 10.17: Any two minimal fibre spaces which are deformation retracts of a given fibre space of Kan complexes are isomorphic.

CONVENTIONS 10.18: By a minimal sub fibration $p: E' \to B$ of a Kan fibration $p: E \to B$ we will understand one such that (E', p, B) is a strong deformation retract of (E, p, B). Similarly, a minimal sub fibre space (E', p', B') of a fibre space (E, p, B) of Kan complexes will mean one which is a strong deformation retract.

§11. Fibre products and fibre bundles

In this section we define fibre products and begin the study of fibre bundles. In particular, we will prove that every minimal fibration is a fibre bundle.

DEFINITION 11.1: Let $p\colon\ E \to B$ and $f\colon\ A \to B$ be maps. Define $E^f = \{(e, a) | p(e) = f(a)\} \subset E \times A$ and let $p^f\colon\ E^f \to A$ and $\tilde{f}\colon\ E^f \to E$ be given by $p^f(e, a) = a$ and $\tilde{f}(e, a) = e$. The commutative diagram

$$\begin{array}{ccc} E^f & \xrightarrow{\ \tilde{f}\ } & E \\ \big\downarrow{\scriptstyle p^f} & & \big\downarrow{\scriptstyle p} \\ A & \xrightarrow{\ f\ } & B \end{array}$$

is called the fibre product of p and f. p^f is called the map induced from p by f.

The following lemma results immediately from the definition.

LEMMA 11.2: If p is a Kan fibration, so is p^f. If p is a minimal fibration, so is p^f.

If p is a Kan fibration, then (E^f, p^f, A) is called the fibre space induced from (E, p, B) by f. Fibre products satisfy the following universal property.

PROPOSITION 11.3: Let

$$\begin{array}{ccc} E' & \xrightarrow{\ g\ } & E \\ \big\downarrow{\scriptstyle p'} & & \big\downarrow{\scriptstyle p} \\ B' & \xrightarrow{\ f\ } & B \end{array}$$

be a commutative diagram. Then there exists a unique map

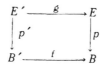

$h\colon E' \to E^f$ such that $g = \tilde{f} \circ h$ and $p' = p^f \circ h$.

Proof: The result is obvious: we must define

$$h(e') = (g(e'), p'(e')).$$

PROPOSITION 11.4: Let $(g, f)\colon (E', p', B') \to (E, p, B)$ be a map of fibre spaces. Suppose that p' is onto and that for $b' \in B'_q$, $g|p'^{-1}(b')$ is a one-to-one correspondence onto $p^{-1}(f(b'))$. Then the map $h\colon E' \to E^f$ such that $g = \tilde{f} \circ h$ and $p' = p^f \circ h$ is an isomorphism.

Proof: Define $j\colon E^f \to E'$ as follows. Suppose $(e, b') \in E^f_q$. Choose any $x \in E'_q$ such that $p'(x) = b'$. Then

$$pg(x) = f(b') = p(e).$$

By assumption, there exists a unique $e' \in E'_q$ such that $g(e') = e$ and $p'(e') = p'(x) = b'$. Define $j(e, b') = e'$. Then $j \circ h = 1$ and $h \circ j = 1$.

Next we show that homotopic maps induce fibre spaces of the same homotopy type.

PROPOSITION 11.5: Let $p\colon E \to B$ be a Kan fibration and let $f \simeq g\colon A \to B$. Then there exists a strong homotopy equivalence between (E^f, p^f, A) and (E^g, p^g, A).

Proof: Let $F\colon A \times I \to B$ be the given homotopy and let $(E^*, p^*, A \times I)$ denote the fibre space induced by F. Identify E^f with $p^{*-1}(A \times (0)) \subset E^*$ and let $x\colon E^f \subset E^*$. Identify E^g with $p^{*-1}(A \times (1)) \subset E^*$ and let $x'\colon E^g \subset E^*$. For notational simplicity, let $q\colon K \to L$ denote the Kan fibration $(p^*)^{E^f}\colon (E^*)^{E^f} \to (A \times I)^{E^f}$ and form $q'\colon K' \to L'$ using E^g instead of E^f. Now $p^f \times 1 \in L_1$ and $\partial_1(p^f \times 1) = q(x)$, hence there exists $y \in K_1$ such that $q(y) = p^f \times 1$ and $\partial_1 y = x$. Then $\partial_0 y$ is a map $E^f \to E^g$. Similarly, there exists $y' \in K'_1$ such that $g'(y') = p^g \times 1$ and

$\partial_0 y' = x'$. $\partial_1 y'$ is a map $E^g \to E^f$. Again, since q is a Kan fibration, there exists $z \in K_2$ such that $q(z) = s_0(p^f \times 1)$, $\partial_0 z = y$ and $\partial_1 z = y' \circ (\partial_0 y \times 1)$. Let $w = \partial_2 z$ Then

$$\partial_0 w = x \text{ and } \partial_1 w = \partial_1 y' \circ \partial_0 y .$$

$q(w) = s_0 \partial_1 (p^f \times 1) = s_0 q(x)$ and therefore the image of w is contained in E^f and $p^f \circ w(e, k) = p^f(e)$, $e \in E^f_q$, $k \in I_q$. Thus w gives a strong homotopy $\partial_1 y' \circ \partial_0 y \simeq 1$ in E^f. Similarly, we may obtain a strong homotopy $\partial_0 y \circ \partial_1 y' \simeq 1$ in E^g, and this implies the result.

COROLLARY 11.6: Let p: $E \to B$ be a minimal fibration and let $f \simeq g$: $A \to B$. Then there exists a strong isomorphism between (E^f, p^f, A) and (E^g, p^g, A).

Proof: This follows from Lemma 11.2, Proposition 10.13, and the result above.

COROLLARY 11.7: Let $(F, \psi) \to (E, \psi) \to (B, \phi)$ be a fibre sequence and suppose that B is contractible. Then there is a strong homotopy equivalence $(g, 1)$: $(E, p, B) \to (F \times B, p^*, B)$ where $p^*(f, b) = b$. If p is minimal, $(g, 1)$ is a strong isomorphism.

Proof: If j: $B \to \phi$, then by assumption $1 \simeq j$: $B \to B$. Since $F \times B = E^j$ and $E^1 = E$, the result follows.

At this point, we are ready to define fibre bundles.

DEFINITION 11.8: A map p: $E \to B$ is called a fibre bundle if p is onto and if for every map $\overline{b}$: $\Delta[n] \to B$, the induced map $p^{\overline{b}}$: $E^{\overline{b}} \to \Delta[n]$ is strongly isomorphic to p^*: $F \times \Delta[n] \to \Delta[n]$, where $p^*(f, k) = k$ and where F is a given complex called the fibre of the bundle. If F is a Kan complex, then p: $E \to B$ is called a Kan fibre bundle.

LEMMA 11.9: A Kan fibre bundle is a Kan fibration.

Proof: Let $x_0, \ldots, x_{k-1}, x_{k+1}, \ldots, x_{q+1}$ be $q+1$ compatible q-simplices of E and suppose $\partial_i b = x_i$, $i \neq k$, $b \in B_{q+1}$. Let $\alpha(b)\colon F \times \Delta[q+1] \to E^{\bar{b}}$ be a strong isomorphism, say

$$\alpha(b)(y_i, \partial_i \Delta_{q+1}) = (x_i, \partial_i \Delta_{q+1}) .$$

There exists $y \in F_{q+1}$ such that $\partial_i y = y_i$, $i \neq k$, and if

$$\alpha(b)(y, \Delta_{q+1}) = (x, \Delta_{q+1}) ,$$

then $p(x) = b$ and $\partial_i x = x_i$, $i \neq k$.

LEMMA 11.10: $\Delta[n]$ is contractible for all n.

Proof: Define $h_0(\Delta_n) = s_0 \Delta_n$, $h_1(\Delta_n) = s_0 \Delta_n$, and $h_i(\Delta_n) = s_{i-1} \cdots s_0 \partial_1 \cdots \partial_{i-1} \Delta_n$, $1 < i \leq n$. Then if $j\colon \Delta[n] \to (0) \subset \Delta[n]$, it is easily verified that $h\colon 1 \simeq j$.

THEOREM 11.11: Every minimal fibration with a connected base complex is a Kan fibre bundle.

Proof: Let $p\colon E \to B$ be the given minimal fibration and let $\bar{b}\colon \Delta[n] \to B$. By Lemmas 11.2 and 11.10, $p^{\bar{b}}$ is minimal and $\Delta[n]$ is contractible, hence by Corollary 11.7 $(E^{p^{\bar{b}}}, p^{\bar{b}}, \Delta[n])$ is strongly isomorphic to a product $(F_b \times \Delta[n], p^*, \Delta[n])$. $\bar{b}$ is homotopic to a map $\bar{\phi}\colon \Delta[n] \to B$, where ϕ is a vertex of B, and thus by Corollary 11.6 $(F_b \times \Delta[n], p^*, \Delta[n])$ is strongly isomorphic to $(F_\phi \times \Delta[n], p^*, \Delta[n])$. Now F_ϕ is the fibre over ϕ and since, by Corollary 7.11, any two fibres are isomorphic, the result follows.

COROLLARY 11.12: Every Kan fibration with a connected base complex contains a Kan fibre bundle as a strong deformation retract.

§12. Weak homotopy type

In this section we will use the earlier material of this chapter to find sufficient conditions for Kan complexes and Kan fibrations to be of the same homotopy type.

We note first the following corollary of Lemma 10.3.

LEMMA 12.1: If K is a minimal complex, then each $(K^{(n)}, p, K^{(m)})$ is a minimal fibre space.

Since a connected minimal complex K has only one vertex ϕ, we will denote $\pi_q(K, \phi)$ by $\pi_q(K)$.

LEMMA 12.2: Let K be a connected minimal complex. Consider the fibre space $(K, p, K^{(q-1)})$, $q \geq 1$. If $b \in K_q^{(q-1)}$, let

$$H_b = \{x \mid p(x) = b\},$$

that is, if $x \in H_b$, then $y \in H_b$ if and only if $\partial_i y = \partial_i x$ for all i. Then the homotopy classes of elements of H_b are in one to one correspondence with the elements of $\pi_q(K)$.

Proof: By Lemma 12.1 and Theorem 11.1, $p\colon K \to K^{(q-1)}$ is a Kan fibre bundle. Let ϕ denote the vertex of $K^{(q-1)}$. Define $\bar{b}$ and $\bar{\phi}\colon \Delta[q] \to K^{(q-1)}$ in the obvious way. Then $(E^{\bar{b}}, p^{\bar{b}}, \Delta[q])$ is strongly isomorphic to $(E_q(K) \times \Delta[q], p^*, \Delta[n])$. where $E_q(K)$ is the q-th Eilenberg subcomplex of K. Therefore the sets of homotopy classes of elements of

$$\{(x, \Delta_q) \mid x \in H_b\} \subset E^{\bar{b}}$$

and of elements of $\{(y, \Delta_q) \mid y \in H_\phi\} \subset E_q(K) \times \Delta[q]$ are in one-to-one correspondence. Noting that $H_\phi = \tilde{E}_q(K)_q$, the latter set is $\pi_q(E_q(K)) \cong \pi_q(K)$. Since the former set is that of homotopy classes of elements of H_b, the result is proven.

DEFINITION 12.3: A map $f\colon K \to L$ is called a weak (or singular) homotopy equivalence if $f_*\colon \pi_q(K, \phi) \to \pi_q(L, f(\phi))$ is an isomorphism for all q.

THEOREM 12.4: Let K and L be connected minimal complexes. Then a weak homotopy equivalence $f\colon K \to L$ is an isomorphism.

Proof: Let $(K^{(n+1)}, p, K^{(n)})$ and $(L^{(n+1)}, p', L^{(n)})$ be the n-th terms in the natural Postnikov systems for K and L. By Lemma 12.1, both of these are minimal fibre spaces. Since

$$K_q = K_q^{(n)} \quad \text{if } q \le n,$$

it suffices to prove that $f^{(n)}\colon K^{(n)} \to L^{(n)}$ is an isomorphism for all n. $f^{(0)}\colon K^{(0)} \xrightarrow{\cong} L^{(0)}$ is clear, so assume that

$$f^{(n-1)}\colon K^{(n-1)} \xrightarrow{\cong} L^{(n-1)}, \quad n \ge 0.$$

Consider the commutative diagram

$$\begin{array}{ccc} K^{(n)} & \xrightarrow{f^{(n)}} & L^{(n)} \\ \downarrow{\scriptstyle p} & & \downarrow{\scriptstyle p'} \\ K^{(n-1)} & \xrightarrow{f^{(n-1)}} & L^{(n-1)} \end{array}$$

(i) Suppose $f^{(n)}(x) = f^{(n)}(y)$, $x, y \in K_q^{(n)}$. We must prove that $x = y$. $f^{(n-1)}p(x) = f^{(n-1)}p(y)$, hence $p(x) = p(y)$. This implies $x = y$ if $q \le n-1$. Suppose $q = n$. Then by Lemma 12.2, the homotopy classes $[x]$ and $[y]$ can be taken to be elements of $\pi_n(K) = \pi_n(K^{(n)})$, hence by naturality $f_*[x] = f_*[y]$. Since f_* is an isomorphism, this implies $[x] = [y]$, that is, $x \sim y$. Then by minimality $x = y$. Assume the result for $q' < q$, $q > n$. Then $\partial_i x = \partial_i y$ for all i and since $\pi_q(K^{(n)}) = 0$, $x = y$ follows from Lemma 12.2.

(ii) Let $y \in L_q^{(n)}$. We must find $x \in K_q^{(n)}$ such that $f^{(n)}(x) = y$. If $q \le n-1$, the result is obvious. Suppose $q = n$. $[y]$ can be taken to be an element of $\pi_n(L) = \pi_n(L^{(n)})$. Since f_* is an isomorphism, there exists $x \in K_n$ such that $f_*[x] = [y]$. Then $f^{(n)}(x) \sim y$, hence $f^{(n)}(x) = y$. Now assume the result for $q' < q$, $q > n$. Let $f^{(n)}(x_i) = \partial_i y$. By the extension condition, there exists $x \in K_q^{(n)}$ such that $\partial_i x = x_i$, $i < q$. Then

$$\partial_i f^{(n)}(x) = \partial_i y, \qquad i < q,$$

hence by minimality $\partial_q f^{(n)}(x) = \partial_q y$. Now since $\pi_q(L^{(n)}) = 0$, it follows from Lemma 12.2 that $f^{(n)}(x) = y$.

The following theorem is an immediate corollary.

THEOREM 12.5: Let K and L be connected Kan complexes. Then the following conditions are equivalent.

(i) K and L have the same homotopy type.

(ii) There exists a weak homotopy equivalence $f\colon K \to L$.

(iii) K and L have isomorphic minimal subcomplexes.

Now we outline the parallel development for Kan fibrations.

LEMMA 12.6: Let $p\colon E \to B$ be a minimal fibration, where E, B, and the fibre F are connected Kan complexes. Then each of the Kan fibrations $p'\colon E^{(n)} \to B$ and $p'\colon E^{(n)} \to E^{(m)}$ is minimal.

Proof: Let $u, v \in E_q$ represent $x, y \in E_q^{(n)}$. Suppose

$$\partial_i x = \partial_i y, \quad i \neq k, \quad \text{and} \quad p'(x) = p'(y).$$

The latter implies $p(u) = p(v)$. If $q \leq n+1$, $\partial_i x = \partial_i y$ implies $\partial_i u = \partial_i v$ and, by the minimality of p, $\partial_k u = \partial_k v$. If $q > n+1$, $\partial_i \partial_k x = \partial_i \partial_k y$ implies $\overline{\partial_k u} = \overline{\partial_k v}$ on $\Delta[q-1]^n$, hence $\partial_k x = \partial_k y$.

DEFINITION 12.7: A map $(\tilde{f}, f)\colon (E', p', B') \to (E, p, B)$ of fibre spaces of Kan complexes is called a weak homotopy equivalence if it induces an isomorphism of the homotopy exact sequence of (E', p', B') onto that of (E, p, B).

THEOREM 12.8: Let (E', p', B') and (E, p, B) be minimal fibre spaces of connected Kan complexes with connected fibres. Then a weak homotopy equivalence $(\tilde{f}, f)\colon (E', p', B') \to (E, p, B)$ is an isomorphism.

THEOREM 12.9: Let (E', p', B') and (E, p, B) be fibre spaces of connected Kan complexes with connected fibres. Then the following are equivalent:

(i) (E', p', B') and (E, p, B) have the same homotopy type.

(ii) There exists a weak homotopy equivalence

$$(\tilde{f}, f): (E', p', B') \to (E, p, B)$$

(iii) (E', p', B') and (E, p, B) have isomorphic minimal sub-fibre spaces.

§13. The Hurewicz theorems

In this section, we obtain some of the classical results comparing homotopy groups with homology groups. Throughout this section, homology means homology with integer coefficients.

PROPOSITION 13.1: Let K be a complex. Then $H_0(K) = F(\pi_0(K))$, the free Abelian group generated by $\pi_0(K)$.

Proof: The map $K_0 \to K_0/(\sim)$ of sets induces an epimorphism $C_0(K) = Z_0(K) = F(K_0) \to F(\pi_0(K))$. Clearly if $x, y \in K_0$, then $x \sim y$ if and only if there exists a 1-chain w such that $\partial(w) = x - y$. This implies the result.

From now on, we consider the reduced homology groups $\tilde{H}_n(K) = H_n(\tilde{C}(K))$ of a pair (K, ϕ), Here $\tilde{C}(K) = C(K)/C(\phi)$. Clearly $\tilde{H}_n(K) = H_n(K)$, $n > 0$, and $\tilde{H}_0(K)$ has one less free generator than $H_0(K)$.

DEFINITIONS 13.2: Let (K, L, ϕ) be a Kan triple. Suppose x represents $\alpha \in \pi_n(K, \phi)$, and consider x as a cycle of $\tilde{C}_n(K)$. Define $h: \pi_n(K, \phi) \to \tilde{H}_n(K)$ by $h(\alpha) = \{x\}$, where $\{x\}$ denotes the homology class of x. Similarly, suppose y represents $\beta \in \pi_n(K, L, \phi)$.

Let $C(K, L) = C(K)/C(L)$, so that $H_n(C(K, L)) = H_n(K, L)$. $\partial(\overline{y}) = 0$, where $\overline{y}$ denotes the image of y in $C_n(K, L)$, and we define $h\colon \pi_n(K, L, \phi) \to H_n(K, L)$ by $h(\beta) = \{y\}$. The maps h are called the Hurewicz homomorphisms.

LEMMA 13.3. The maps h are well-defined and are homomorphisms of groups.

Proof: We give the proof for the absolute case, that for the relative case being similar. Suppose $x, y \in \tilde{K}_n$ and $z\colon x \sim y$. Then $\partial(z) = (-1)^n(x-y)$, hence h is well-defined. Suppose next that x, y represent $\alpha, \beta \in \pi_n(K, \phi)$ and $w \in K_{n+1}$ satisfies $\partial_i w = \phi$, $0 \leq i < n-1$, $\partial_{n-1} w = x$, and $\partial_{n+1} w = y$, so that $\alpha\beta = [\partial_n w]$. $\partial(w) = (-1)^{n+1}(x - \partial_n w + y)$ and therefore

$$h(\alpha\beta) = h(\alpha) + h(\beta).$$

PROPOSITION 13.4: The Hurewicz homomorphisms define natural transformations of functors. If (K, L, ϕ) is a Kan triple, then the following is a commutative diagram of exact sequences:

$$\begin{array}{ccccccccc}
\cdots \to & \pi_{n+1}(K, L, \phi) & \xrightarrow{\partial} & \pi_n(L, \phi) & \xrightarrow{i} & \pi_n(K, \phi) & \xrightarrow{j} & \pi_n(K, L, \phi) & \to \cdots \\
 & \downarrow h & & \downarrow h & & \downarrow h & & \downarrow h & \\
\cdots \to & H_{n+1}(K, L) & \xrightarrow{\partial} & \tilde{H}_n(L) & \xrightarrow{i} & \tilde{H}_n(K) & \xrightarrow{j} & H_n(K, L) & \to \cdots
\end{array}$$

Proof: The result is clear from the definitions, recalling that $\partial\colon H_{n+1}(K, L) \to \tilde{H}_n(L)$ is defined by $\partial\{\overline{y}\} = \{\partial(y)\}$, $\overline{y} \in Z_n(K, L)$.

THEOREM 13.5 (Poincaré): Let (K, ϕ) be a connected Kan pair. Then the map $h'\colon \pi_1(K, \phi)/[\pi_1(K, \phi), \pi_1(K, \phi)] \to H_1(K)$ induced by h is an isomorphism.

Proof: We may assume $K = E_1(K, \phi)$. Then

$$\tilde{C}_1(K) = \tilde{Z}_1(K) = F(K_1 - \phi).$$

Let $j\colon \tilde{Z}_1(K) \to \pi_1(K)/[\pi_1(K), \pi_1(K)]$ be the natural epimorphism induced from $K_1 \to K_1/(\sim)$. If $x \in \tilde{C}_2(K)$, then

$$\partial(x) = \partial_0 x - \partial_1 x + \partial_2 x$$

and by definition of the group structure in $\pi_1(K)$, $j\partial(x) = 0$. Thus j induces $j'\colon H_1(K) \to \pi_1(K)/[\pi_1(K), \pi_1(K)]$. Clearly $j' \circ h' = 1$ and $h' \circ j' = 1$, and this completes the proof.

THEOREM 13.6 (Hurewicz): Let (K, ϕ) be an $(n-1)$-connected Kan pair, $n \geq 2$. Then $H_i(K) = 0$, $0 < i \leq n-1$, and $h\colon \pi_n(K) \cong H_n(K)$.

Proof: We may assume $K = E_n(K, \phi)$. Then K has only one simplex in each dimension $< n$, hence $H_i(K) = 0$, $0 < i \leq n-1$, and $\tilde{C}_n(K) = \tilde{Z}_n(K) = F(K_n - \phi)$. $K_n \to K_n/(\sim)$ induces a natural epimorphism $j\colon \tilde{Z}_n(K) \to \pi_n(K)$. As in the proof above, it suffices to prove $j\partial(x) = 0$, $x \in K_{n+1}$. If $n = 2$, this follows immediately from step (iii) of the proof of Proposition 4.4 on page 11; if $n > 2$, a tedious but essentially similar computation gives the result.

COROLLARY 13.7: If K is a 1-connected Kan complex and

$$H_i(K) = 0 \text{ for all } i > 0,$$

then K is contractible.

Similarly, we can prove

THEOREM 13.8: Let (K, L, ϕ) be a Kan triple, where K and L are 1-connected Kan complexes. Then (K, L, ϕ) is $(n-1)$-connected if and only if $H_i(K, L) = 0$, $0 \leq i \leq n-1$, and in that case

$$h\colon \pi_n(K, L, \phi) \to H_n(K, L)$$

is an isomorphism.

The hypothesis that f is an inclusion can be eliminated in the following result, but some technical work, or use of geometric realization, is required since the natural simplicial mapping cylinder of a map of Kan complexes need not itself be a Kan complex.

THEOREM 13.9 (Whitehead): Let $f: K \to L$ be an inclusion of 1-connected Kan complexes and let $n \geq 2$ be an integer. Then the following are equivalent:

(i) f_*: $\pi_i(K, \phi) \to \pi_i(L, f(\phi))$ is an isomorphism for $i < n$ and an epimorphism for $i = n$.

(ii) f_*: $H_i(K) \to H_i(L)$ is an isomorphism for $i < n$ and an epimorphism for $i = n$.

Proof: Since we have assumed that f is an inclusion, this follows immediately from the theorem above.

BIBLIOGRAPHICAL NOTES ON CHAPTER II

The concept of simplicial fibre space is due to Kan [25], [30]. Complete proofs of Theorems 7.8, 7.13 and 7.16 may be found in Cartan [6], in Gugenheim [18], in Kan [33], and in MacLane [41].

The original definition of Postnikov systems is of course that of Postnikov [54]. An alternative semi-simplicial treatment is that of Heller [21]. We have followed Moore [52],[53]. The fibre spaces $(K, p, K^{(n)})$ are the simplicial analogs of "construction II" of Cartan and Serre [4]. The Eilenberg subcomplexes of the total singular complex of a space were defined by Eilenberg in [10]. Eilenberg-MacLane complexes were introduced and studied in the series of papers [13].

The construction of the minimal subcomplex of the total singular complex of a space is due to Eilenberg and Zilber [15]. Our treatment is essentially that of Moore [52], [53]. The theorem that every minimal fibration is a Kan fibre bundle is due to Barratt, Gugenheim, and Moore [1]. The equivalence of weak homotopy type and homotopy type for Kan complexes is of course the analog of a theorem due to Whitehead [62] on CW-complexes.

CHAPTER III

GEOMETRIC REALIZATION

§14. The realization

Throughout this chapter, Δ_n will denote the topological n-simplex, $\Delta_n = \{(t_0,\dots,t_n) \mid 0 \leq t_i \leq 1, \Sigma\, t_i = 1\} \subset R^{n+1}$. We define maps $\delta_i\colon \Delta_{n-1} \to \Delta_n$ and $\sigma_i\colon \Delta_{n+1} \to \Delta_n$ by:

$$\delta_i(t_0,\dots,t_{n-1}) = (t_0,\dots,t_{i-1}, 0, t_i,\dots,t_{n-1}) ,$$

$$\sigma_i(t_0,\dots,t_{n+1}) = (t_0,\dots,t_i + t_{i+1},\dots,t_{n+1}) .$$

A point of Δ_n is called interior if $n = 0$ or if $0 < t_i < 1$ for all i. Note that every $u_n \in \Delta_n$ can be uniquely expressed in the form $u_n = \delta_{i_q} \cdots \delta_{i_1} u_{n-q}$, where $u_{n-q} \in \Delta_{n-q}$ is an interior point and $0 \leq i_1 < \cdots < i_q \leq n$.

Now let K be a complex. Give K the discrete topology and form the disjoint union $\overline{K} = \cup_{n \geq 0}(K_n \times \Delta_n)$. Define an equivalence relation $\approx$ in $\overline{K}$ by:

$$(\partial_i k_n, u_{n-1}) \approx (k_n, \delta_i u_{n-1}),\quad k_n \in K_n,\quad u_{n-1} \in \Delta_{n-1},$$

and

$$(s_i k_n, u_{n+1}) \approx (k_n, \sigma_i u_{n+1}),\quad k_n \in K_n,\quad u_{n+1} \in \Delta_{n+1} .$$

The identification space $T(K) = \overline{K}/(\approx)$ is called the geometric realization of K. The class of (k_n, u_n) in $T(K)$ will be denoted by $|k_n, u_n|$. If $f\colon K \to L$ is a simplicial map, then f induces the continuous map $T(f)\colon T(K) \to T(L)$ defined by

$$T(f)|k_n, u_n| = |f(k_n), u_n| .$$

It is clear that T thus defined is a covariant functor from the category of simplicial sets to that of topological spaces.

We will prove that $T(K)$ is actually a CW-complex. We recall the definition. A cell complex X is a Hausdorff space which is the disjoint union of open cells e^n subject to the requirement that for each cell e^n, if $\bar{e}^n$ denotes the closure of e^n and $\dot{\Delta}_n$ the boundary of Δ_n, then there exists a map $f\colon \Delta_n \to \bar{e}^n$ such that $f|(\Delta_n - \dot{\Delta}_n)$ is a homeomorphism onto e^n and $f(\dot{\Delta}_n)$ is contained in the union of the cells of dimension less than n. A subcomplex Y of X is a union of cells of X such that $e^n \subset Y$ implies $\bar{e}^n \subset Y$. A cell complex is closure finite if each $\bar{e}^n$ is contained in a finite subcomplex and has the weak topology if a subset is closed provided its intersection with each $\bar{e}^n$ is closed. A closure finite cell complex with the weak topology is called a CW-complex.

THEOREM 14.1: $T(K)$ is a CW-complex having one n-cell for each non-degenerate n-simplex of K.

A point (k_n, u_n) of $\bar{K}$ is said to be non-degenerate if $k_n \in K_n$ is non-degenerate and $u_n \in \Delta_n$ is interior. The theorem will follow immediately from the following lemma.

LEMMA 14.2: Every point $(k_n, u_n) \in \bar{K}$ is equivalent to a unique non-degenerate point.

Proof: By formula (3) on page 4, every $k_n \in K_n$ can be uniquely expressed in the form $k_n = s_{j_p} \cdots s_{j_1} k_{n-p}$, where $k_{n-p} \in K_{n-p}$ is non-degenerate and $0 \leq j_1 < \cdots < j_p < n$. The indices j which occur are precisely those such that $k_n \in s_j K_{n-1}$.

Define λ: $\bar{K} \to \bar{K}$ by

$$\lambda(k_n, u_n) = (k_{n-p}, \sigma_{j_1} \dots \sigma_{j_p} u_n), \quad k_n = s_{j_p} \dots s_{j_1} k_{n-p},$$
$$k_{n-p} \text{ non-degenerate}, \qquad 0 \le j_1 < \dots < j_p < n.$$

Similarly, define ρ: $\bar{K} \to \bar{K}$ by

$$\rho(k_n, u_n) = (\partial_{i_1} \dots \partial_{i_q} k_n, u_{n-q}), \quad u_n = \delta_{i_q} \dots \delta_{i_1} u_{n-q},$$
$$u_{n-q} \text{ interior}, \qquad 0 \le i_1 < \dots < i_q \le n.$$

The composition $\lambda \circ \rho$ carries each point into an equivalent non-degenerate point. Uniqueness follows since $x \approx x'$ implies $\lambda \circ \rho(x) = \lambda \circ \rho(x')$.

Next we prove that, under mild hypotheses,

$$T(K \times L) = T(K) \times T(L).$$

Let π: $K \times L \to K$ and π': $K \times L \to L$ be the projections and define η: $T(K \times L) \to T(K) \times T(L)$ by $\eta = T(\pi) \times T(\pi')$.

THEOREM 14.3: η: $T(K \times L) \to T(K) \times T(L)$ is one-to-one and onto. If $T(K) \times T(L)$ is a CW-complex, then η is a homeomorphism.

Proof: If $z \in T(K \times L)$ has non-degenerate representative $(k_n \times \ell_n, w_n)$, then $T(\pi)(z) = |k_n, w_n|$ has non-degenerate representative $\lambda(k_n, w_n)$ and $T(\pi')(z) = |\ell_n, w_n|$ has non-degenerate representative $\lambda(\ell_n, w_n)$, where λ is as defined in the proof of Lemma 14.2. We define an inverse function $\bar{\eta}$: $T(K) \times T(L) \to T(K \times L)$ as follows. Let $(x, y) \in T(K) \times T(L)$, where x and y have non-degenerate representatives (k_a, u_a) and (ℓ_b, v_b). If $u_a = (t_0, \dots, t_a)$ and $v_b = (t_0', \dots, t_b')$, define

$$u^m = \sum_{i=0}^{m} t_i \quad \text{and} \quad v^n = \sum_{j=0}^{n} t_j' \quad .$$

Let $r_0 < \dots < r_c = 1$ be the sequence obtained by arranging the distinct elements of $\{u^m\} \cup \{v^n\}$ in order of magnitude, and define

$t_i'' = r_i - r_{i-1}$, $0 \le i \le c$, $r_{-1} = 0$. Clearly $0 < t_i'' < 1$ and

$$\sum_{i=0}^{c} t_i'' = r_c = 1,$$

hence $w_c = (t_0'', \ldots, t_c'')$ is an interior point of Δ_c. Let $i_1 < \cdots < i_{c-a}$ be those integers i such that $r_i \notin \{u^m\}$ and let $j_1 < \cdots < j_{c-b}$ be those integers j such that $r_j \notin \{v^n\}$. Then $\{i_\alpha\}$ and $\{j_\beta\}$ are disjoint, $u_a = \sigma_{i_1} \cdots \sigma_{i_{c-a}} w_c$, and $v_b = \sigma_{j_1} \cdots \sigma_{j_{c-b}} w_c$. Define

$$\bar{\eta}(x, y) = |(s_{i_{c-a}} \cdots s_{i_1} k_a) \times (s_{j_{c-b}} \cdots s_{j_1} k_b), w_c| .$$

Clearly

$$T(\pi)\bar{\eta}(x, y) = |\lambda(s_{i_{c-a}} \cdots s_{i_1} k_a, w_c)| = |k_a, u_a| = x$$

and $T(\pi')\bar{\eta}(x, y) = y$, so that $\eta \circ \bar{\eta} = 1$. Taking z as above,

$$\bar{\eta} \circ \eta(z) = \bar{\eta}(|\lambda(k_n, w_n)|, |\lambda(\ell_n, w_n)|) = |(k_n \times \ell_n, w_n)| = z .$$

Finally, we observe that $\bar{\eta}$ is continuous on each product cell of $T(K) \times T(L)$ and, if $T(K) \times T(L)$ is a CW-complex, this implies that $\bar{\eta}$ is continuous, hence that η is a homeomorphism.

REMARK 14.4: The proof above parallels that for simplicial complexes given in [14, p. 68]. The hypothesis that $T(K) \times T(L)$ is a CW-complex holds if K and L are both countable [47, p. 272] or if either $T(K)$ or $T(L)$ is locally finite (i. e., every point is an inner point of a finite subcomplex) [62, p. 227].

COROLLARY 14.5: A simplicial homotopy $F: K \times I \to L$ induces a topological homotopy $T(F) \circ \bar{\eta}: T(K) \times T(I) \to T(L)$.

COROLLARY 14.6: If K is a countable simplicial monoid, group, or Abelian group, then $T(K)$ is a topological monoid, group, or Abelian group.

§15. Adjoint functors

In this section, we define adjoint functors and develop a few of their properties.

DEFINITION 15.1: Let $T\colon \mathfrak{A} \to \mathfrak{B}$ and $S\colon \mathfrak{B} \to \mathfrak{A}$ be covariant functors. We say that S and T are adjoint, or that T is an adjoint of S and S a coadjoint of T, if there exists a natural equivalence of functors $\phi\colon \mathrm{Hom}_{\mathfrak{A}}(A, S(B)) \to \mathrm{Hom}_{\mathfrak{B}}(T(A), B)$ (where each side is considered as a functor $\mathfrak{A}^* \times \mathfrak{B} \to \mathcal{C}$, $\mathcal{C}$ the category of sets).

We notice first the following

LEMMA 15.2: Let $S\colon \mathfrak{B} \to \mathfrak{A}$ and $T\colon \mathfrak{A} \to \mathfrak{B}$ be covariant functors. Then

(i) The correspondence between natural transformations $\phi\colon \mathrm{Hom}_{\mathfrak{A}}(A, S(B)) \to \mathrm{Hom}_{\mathfrak{B}}(T(A), B)$ and $\Phi\colon TS \to 1_{\mathfrak{B}}$ obtained by letting ϕ correspond to Φ if $\Phi(B) = \phi(1_{S(B)})$ and

$$\phi(f) = \phi(B) \circ T(f)$$

for $f \in \mathrm{Hom}_{\mathfrak{A}}(A, S(B))$ is one-to-one.

(ii) The correspondence between natural transformations $\psi\colon \mathrm{Hom}_{\mathfrak{B}}(T(A), B) \to \mathrm{Hom}_{\mathfrak{A}}(A, S(B))$ and $\Psi\colon 1_{\mathfrak{A}} \to ST$ obtained by letting ψ correspond to Ψ if $\Psi(A) = \psi(1_{T(A)})$ and

$$\psi(g) = S(g) \circ \Psi(A)$$

for $g \in \mathrm{Hom}_{\mathfrak{B}}(T(A), B)$ is one-to-one.

PROPOSITION 15.3: Let $\phi\colon \mathrm{Hom}_{\mathfrak{A}}(A, S(B)) \to \mathrm{Hom}_{\mathfrak{B}}(T(A), B)$ correspond to $\Phi\colon TS \to 1_{\mathfrak{B}}$. Let

$$\psi\colon \mathrm{Hom}_{\mathfrak{B}}(T(A), B) \to \mathrm{Hom}_{\mathfrak{A}}(A, S(B))$$

correspond to $\Psi\colon 1_{\mathfrak{A}} \to ST$. Then

(i) $\psi \circ \phi = 1$ if and only if $S \xrightarrow{\Psi S} STS \xrightarrow{S\Phi} S$ is the identity natural transformation of S into S.

(ii) $\phi \circ \psi = 1$ if and only if $T \xrightarrow{T\Psi} TST \xrightarrow{\Phi T} T$ is the identity natural transformation of T into T.

Proof: We prove (i), the proof of (ii) being similar. Let $f \in \mathrm{Hom}_{\mathfrak{A}}(A, S(B))$. The following diagram is commutative:

$$\psi \circ \phi(f) : A \to S(B)$$

$$\begin{array}{ccccc} A & \xrightarrow{\Psi(A)} & ST(A) & \xrightarrow{S\phi(f)} & S(B) \\ \Big\downarrow f & & \Big\downarrow ST(f) & \nearrow S\Phi(B) & \\ S(B) & \xrightarrow{\Psi S(B)} & STS(B) & & \end{array}$$

It follows immediately that $S\Phi(B) \circ \Psi S(B) = 1_{S(B)}$ implies $\psi \circ \phi(f) = f$ and, taking $f = 1_{S(B)}$, that $\psi \circ \phi(f) = f$ implies $S\Phi(B) \circ \Psi S(B) = 1_{S(B)}$.

THEOREM 15.4: Let S, S': $\mathfrak{B} \to \mathfrak{A}$ and T, T': $\mathfrak{A} \to \mathfrak{B}$ be covariant functors. Let ϕ: $\mathrm{Hom}_{\mathfrak{A}}(A, S(B)) \to \mathrm{Hom}_{\mathfrak{B}}(T(A), B)$ be a natural equivalence with inverse ψ and let

$$\phi'\colon \mathrm{Hom}_{\mathfrak{A}}(A, S'(B)) \to \mathrm{Hom}_{\mathfrak{B}}(T'(A), B)$$

be a natural equivalence with inverse ψ'. Suppose that τ: $T' \to T$ (resp. σ: $S \to S'$) is a natural transformation. Then there exists a unique natural transformation σ: $S \to S'$ (resp. τ: $T' \to T$) such that the following is a commutative diagram:

$$(*) \qquad \begin{array}{ccc} \mathrm{Hom}_{\mathfrak{B}}(T(A), B) & \xrightarrow{\psi} & \mathrm{Hom}_{\mathfrak{A}}(A, S(B)) \\ \Big\downarrow \mathrm{Hom}_{\mathfrak{B}}(\tau(A), B) & & \Big\downarrow \mathrm{Hom}_{\mathfrak{A}}(A, \sigma(B)) \\ \mathrm{Hom}_{\mathfrak{B}}(T'(A), B) & \xrightarrow{\psi'} & \mathrm{Hom}_{\mathfrak{A}}(A, S'(B)) \end{array} .$$

If τ (resp. σ) is a natural equivalence, then so is σ (resp. τ).

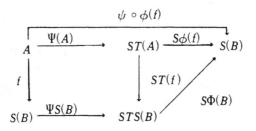

Proof: Given $f \in \mathrm{Hom}_{\mathcal{Q}}(A, S(B))$, the commutativity of the diagram (*) is seen to imply $\sigma(B) \circ f = \psi'(\phi(f) \circ \tau(A))$. In particular, taking $A = S(B)$ and $f = 1_{S(B)}$, we must have

$$\text{(i)} \qquad \sigma(B) = \psi'(\phi(1_{S(B)}) \circ \tau S(B)) = \psi'(\Phi(B) \circ \tau S(B)) .$$

Thus given τ, σ must be defined by (i). It is easily verified that σ so defined is a natural transformation. Similarly, for

$$g \in \mathrm{Hom}_{\mathcal{B}}(T(A), B),$$

the commutativity of the diagram (*) implies

$$g \circ \tau(A) = \phi'(\sigma(B) \circ \psi(g)) .$$

Taking $B = T(A)$ and $g = 1_{T(A)}$, we must have

$$\text{(ii)} \qquad \tau(A) = \phi'(\sigma T(A) \circ \psi(1_{T(A)})) = \phi'(\sigma T(A) \circ \Psi(A)) .$$

Given σ, τ must be defined by (ii) and naturality is easily verified. The last statement follows from the uniqueness: if τ has inverse τ^{-1}, then if σ' is induced from τ^{-1}, $\sigma\sigma'$ and $\sigma'\sigma$ are induced from $\tau\tau^{-1}$ and $\tau^{-1}\tau$, hence $\sigma\sigma'$ and $\sigma'\sigma$ are both indentities.

Note that the previous theorem implies that any two adjoints of a functor S are naturally equivalent and any two coadjoints of a functor T are naturally equivalent.

§16. Comparison of simplicial sets and topological spaces

Let $\mathcal{S}$ denote the category of simplicial sets and $\mathcal{T}$ that of topological spaces. We claim that the realization functor $T\colon \mathcal{S} \to \mathcal{T}$ is adjoint to the total singular complex functor $S\colon \mathcal{T} \to \mathcal{S}$. Thus define $\phi\colon \mathrm{Hom}_{\mathcal{S}}(K, S(X)) \to \mathrm{Hom}_{\mathcal{T}}(T(K), X)$ and

$$\psi\colon \mathrm{Hom}_{\mathcal{T}}(T(K), X) \to \mathrm{Hom}_{\mathcal{S}}(K, S(X))$$

by

(1) $\phi(f)|k_n, u_n| = f(k_n)(u_n)$;

(2) $\psi(g)(k_n)(u_n) = g|k_n, u_n|$.

It is easily verified that $\phi \circ \psi = 1$ and $\psi \circ \phi = 1$. The corresponding natural transformations $\Phi\colon TS \to 1_{\mathcal{T}}$ and $\Psi\colon 1_{\mathcal{S}} \to ST$ are given by

(3) $\Phi(X)|k_n, u_n| = k_n(u_n)$;

(4) $\Psi(K)(k_n)(u_n) = |k_n, u_n|$.

Note that $\Psi(K)\colon K \to ST(K)$ is an inclusion of complexes and $\Phi(X)\colon TS(X) \to X$ is a surjection of spaces. By Proposition 15.3, we find

(5) $(S\Phi \circ \Psi S)(X)\colon S(X) \to S(X)$ is the identity map;

(6) $(\Phi T \circ T\Psi)(K)\colon T(K) \to T(K)$ is the identity map.

We claim further that ϕ and ψ preserve homotopies. If $f \simeq g\colon K \to S(X)$ and if $F\colon K \times I \to S(X)$ defines the homotopy, then by Corollary 14.5, $T(F)\colon T(f) \simeq T(g)$. Since $\phi(F) = \Phi S(X) \circ T(F)$, we find $\phi(F)\colon \phi(f) \simeq \phi(g)$. Similarly, if $f \simeq g\colon T(K) \to X$ and $F\colon T(K) \times T(I) \to X$ defines the homotopy, then $S(F)\colon S(f) \simeq S(g)$. Since $\psi(F) = S(F) \circ \Psi(K)$, we find $\psi(F)\colon \psi(f) \simeq \psi(g)$. Summarizing, we have the

THEOREM 16.1: The natural equivalences ϕ and ψ define S and T as adjoint functors. ϕ and ψ induce one-to-one correspondences between the homotopy classes of simplicial maps $K \to S(X)$ and of continuous maps $T(K) \to X$. In particular, $\pi_n(X, x_0) = \pi_n(S(X), S(x_0))$.

Now we consider the maps of the homology and homotopy groups induced by the natural transformations Φ and Ψ.

PROPOSITION 16.2: Let K be a complex, X a space. Then

(i) $\Psi(K)_*: H_*(K) \to H_*(ST(K)) = H_*(T(K))$ is an isomorphism;

(ii) $\Phi(X)_*: H_*(TS(X)) \to H_*(X)$ is an isomorphism.

Proof: (i) Let $D(K) \subset C(K)$ denote the chain subcomplex generated by all the degenerate simplices of K. We will prove later that $C_N(K) = C(K)/D(K)$ is chain homotopic to $C(K)$. If $K^{(n)}$ denotes the n-skeleton of K, then the inclusions $K^{(0)} \subset K^{(1)} \subset \cdots$ induce a filtration of $C_N(K)$. The resulting spectral sequence $\{E^r\}$ converges to $H_*(K)$ and satisfies $E^1_{p,q} = H_{p+q}(K^{(p)}, K^{(p-1)})$. Clearly $E^1_{p,q} = 0$, $q \neq 0$, and $E^1_{p,o}$ is the free Abelian group generated by the non-degenerate p-simplices of K. Therefore

$$E^2_{p,o} = E^\infty_{p,o} = H_p(K).$$

Again, $T(K^{(n)}) = T(K)^{(n)}$, and the inclusions

$$ST(K^{(0)}) \subset ST(K^{(1)}) \subset \cdots$$

give rise to a spectral sequence $\{\bar{E}^r\}$ which converges to $H_*(T(K))$ and satisfies $\bar{E}^1_{p,q} = H_{p+q}(T(K^{(p)}), T(K^{(p-1)}))$. Now the homology of a CW-complex can be computed from the cellular structure, that is, from the subcomplex of the total singular complex generated by the attaching maps, and it follows that the map $E^1 \to \bar{E}^1$ induced by $\Psi(K)$ is an isomorphism. Therefore the spectral sequences are mapped isomorphically, and

$$H_p(K) = E^2_{p,o} \xrightarrow{\simeq} \bar{E}^2_{p,o} = H_p(T(K)),$$

as was to be proven.

(ii) $H_*(TS(X)) = H_*(STS(X))$ and $H_*(X) = H_*(SX)$. By (i), $\Psi S(X)_*: H_*(SX) \to H_*(STS(X))$ is an isomorphism. Since $(S\Phi \circ \Phi S): S(X) \to S(X)$ is the identity, the result follows.

LEMMA 16.3: Let (K, ϕ) be a Kan pair. Then

$$\Psi(K)_*\colon \pi_1(K, \phi) \to \pi_1(ST(K), ST(\phi)) = \pi_1(T(K), T(\phi))$$

is an isomorphism.

Proof: We may assume that K is connected and minimal. Then $\pi_1(K, \phi)$ is a group with one generator for each non-degenerate 1-simplex and one relation for each non-degenerate 2-simplex. $T(K)$ is a CW-complex with one 0-cell and therefore $\pi_1(T(K), T(\phi))$ is a group with one generator for each 1-cell and one relation for each 2-cell. The result follows.

DEFINITIONS 16.4: Let K be a Kan complex having just one vertex ϕ, and let $\pi = \pi_1(K, \phi)$. Define a complex $\tilde{K}$ by $\tilde{K}_n = K_n \times \pi$ with face and degeneracy operators defined on $\tilde{K}_n$ by

(i) $\partial_i(x, a) = (\partial_i x, a), \quad i < n,$

$\partial_n(x, a) = (\partial_n x, [\partial_0^{n-1} x]^{-1} a)$, where ∂_0^{n-1} denotes ∂_0 iterated $n-1$ times, and

(ii) $s_i(x, a) = (s_i x, a)$.

It is easily verified that $\tilde{K}$ is a Kan complex and that the projection $p\colon \tilde{K} \to K$ is a Kan fibration. The fibre F over ϕ satisfies $\pi_n(F, \tilde{\phi}) = 0$, $n > 0$, and $\pi_0(F, \tilde{\phi}) = \pi$, where $\tilde{\phi} = (\phi, e)$, e the identity of π. If x represents $a \in \pi_1(K, \phi)$, then choosing (x, β) such that $\partial_1(x, \beta) = \tilde{\phi}$, that is, such that $[x]^{-1}\beta = e$, we find that $\partial(a) = [\partial_0(x, \beta)] = [(\phi, a)] \in \pi_0(F, \tilde{\phi})$. Therefore

$$\partial\colon \pi_1(K, \phi) \to \pi_0(F, \tilde{\phi})$$

is an isomorphism. By the homotopy exact sequence, $\tilde{K}$ is 1-connected and satisfies $\pi_n(\tilde{K}, \tilde{\phi}) \cong \pi_n(K, \phi)$, $n > 1$. $\tilde{K}$ is called the universal covering complex of K.

REMARKS 16.5: Recall that a map p: $X \to Y$ of topological spaces is called a Serre fibration if p has the covering homotopy property for polyhedra and that this is true if and only if p has this property for each Δ_n. It is easily seen that p is a Serre fibration if and only if $S(p)$ is a Kan fibration. Further, if p: $K \to L$ is a Kan fibration, then it follows from Corollary 7.12 that $T(p)$ is a quasi fibration; J. C. Moore (unpublished) has proven that if p is minimal, then $T(p)$ is a Serre fibration.

THEOREM 16.6: Let (K, ϕ) be a Kan pair and let X be a topological space with base point x_0. Then

(i) $\Psi(K)_*$: $\pi_n(K, \phi) \to \pi_n(ST(K), ST(\phi))$ is an isomorphism for all n, hence $\Psi(K)$ is a homotopy equivalence.

(ii) $\Phi(X)_*$: $\pi_n(TS(X), TS(x_0)) \to \pi_n(X, x_0)$ is an isomorphism for all n, hence $\Phi(X)$ is a homotopy equivalence if X is a CW-complex.

Proof: (i) We may assume that K is connected and minimal. Let $\tilde{K}$ be the universal covering complex of K, p: $\tilde{K} \to K$ the projection, and F the fibre over ϕ. Observe that p is minimal. By Lemma 16.3, Proposition 16.2, and Theorem 13.9

$$\Psi(\tilde{K})_*: \ \pi_n(\tilde{K}, \tilde{\phi}) \cong \pi_n(ST(\tilde{K}), ST(\tilde{\phi}))$$

and trivially $\Psi(F)_*$: $\pi_n(F, \tilde{\phi}) \cong \pi_n(ST(F), ST(\tilde{\phi}))$. By naturality, Ψ induces a map of the homotopy exact sequence of the Kan fibration p to that of the Kan fibration $ST(p)$. and the result follows.

(ii) This follows from (i) and the fact that

$$(S\Phi \circ \Psi S)(X): \ S(X) \to S(X)$$

is the identity.

Finally, we extend the definition of homotopy groups to the category of simplicial sets.

DEFINITION 16.7: Let (K, ϕ) be a pair. Then we define

$$\pi_n(K, \phi) = \pi_n(ST(K), ST(\phi)) ,$$

and similarly for the relative groups.

BIBLIOGRAPHICAL NOTES ON CHAPTER III

The construction of the geometrical realization and the derivation of its properties presented here is due to Milnor [48].

The concept of adjoint functor and the results of section 15 are due to Kan [26]. In [28], Kan has developed a procedure for associating to a complex K a Kan complex $E_x^\infty K$ in such a manner that $T(E_x^\infty K)$ has the same homotopy and homology groups as $T(K)$. This gives an alternative, and self-contained, procedure for extending the definition of homotopy groups to the category of simplicial sets. Another possible definition, also due to Kan, will be presented in Chapter VI.

The CW-hypothesis in Theorem 14.3 and the countability hypothesis in Corollary 14.6 can be eliminated by working in the category of compactly generated spaces.

CHAPTER IV

TWISTED CARTESIAN PRODUCTS AND FIBRE BUNDLES

The concept of twisted Cartesian product will dominate the material to be presented in the next three chapters. We will introduce this notion after obtaining certain special properties of simplicial groups. Then we shall use the notion to study the classification of fibre bundles.

§17. Simplicial groups.

Recall that a simplicial group, or group complex, G is a contravariant functor from Δ^* to the category of groups. Thus each G_n is a group and the face and degeneracy operators are homomorphisms. We let e_n denote the identity of G_n and e the complex consisting of all of the e_n.

THEOREM 17.1: Every group complex G is a Kan complex.

Proof: Suppose $x_0, \ldots, x_{k-1}, x_{k+1}, \ldots, x_{q+1} \in G_q$ satisfy $\partial_i x_j = \partial_{j-1} x_i$, $i < j$, $i, j \neq k$. We must find $x \in G_{q+1}$ such that $\partial_i x = x_i$, $i \neq k$. We first find $u \in G_{q+1}$ such that $\partial_i u = x_i$ if $i < k$. If $k = 0$, this condition is vacuous. Assume $k > 0$. We will find u^r such that $\partial_i u^r = x_i$, $i \leq r$. Let $u^0 = s_0 x_0$; proceeding inductively, suppose u^{r-1} has been defined, $0 < r \leq k-1$. Let $y^{r-1} = s_r((\partial_r u^{r-1})^{-1} x_r)$ and $u^r = u^{r-1} y^{r-1}$. A simple calculation proves that $\partial_i y^{r-1} = e_q$ if $i < r$. It follows that $\partial_i u^r = x_i$ if $i \leq r$, as desired, and we let $u = u^{k-1}$. Next we find $v^r \in G_{q+1}$ such that $\partial_i v^r = x_i$ if $i < k$ or if $i > q-r+1$. Let $v^0 = u$ or $v^0 = e_{q+1}$ if $k = 0$. Proceeding inductively, suppose v^{r-1} has been defined, $0 < r \leq q-k+1$. Let $z^{r-1} = s_{q-r+1}((\partial_{q-r+2} v^{r-1})^{-1} x_{q-r+2})$ and $v^r = v^{r-1} z^{r-1}$. Then $\partial_i z^{r-1} = e_q$ if $i < k$ or if $i > q-r+2$. It follows that $\partial_i v^r = x_i$ if $i < k$ or if $i > q-r+1$, as desired. Finally, letting $x = v^{q-k+1}$, we find $\partial_i x = x_i$, $i \neq k$.

If G is a group complex or a Kan monoid complex, we denote $\pi_n(G, e)$ by $\pi_n(G)$.

PROPOSITION 17.2: Let G be a Kan monoid complex and suppose $x, y \in \tilde{G}_q$, that is, $\partial_i x = e_{q-1} = \partial_i y$ for all i. Then $[x][y] = [xy]$.

Proof: Let $z = s_{q-1}x \cdot s_q y$. Then $\partial_i z = e_{q-1}$, $i < q-1$, $\partial_{q-1} z = x$, $\partial_q z = xy$, and $\partial_{q+1} z = y$. The result follows from the definition of the group structure.

PROPOSITION 17.3: Let G be a Kan monoid complex. Then $\pi_q(G)$ is Abelian, $q \geq 1$.

Proof: Let $w = s_q x \cdot s_{q-1} y$, $x, y \in \tilde{G}_q$. Then $\partial_i w = e_{q-1}$, $i < q-1$, $\partial_{q-1} w = y$, $\partial_q w = xy$, and $\partial_{q+1} w = x$. Therefore $[y][x] = [xy]$. Since $[xy] = [x][y]$ by the previous proposition, the result follows.

There is a useful alternative definition of the homotopy groups of a group complex, which we now develop.

PROPOSITION 17.3: Let G be a group complex and define

$$\overline{G}_q = G_q \cap \operatorname{Ker} \partial_0 \cap \cdots \cap \operatorname{Ker} \partial_{q-1} .$$

Then

(i) $\partial_{q+1}(\overline{G}_{q+1}) \subset \overline{G}_q$, $q \geq 0$.

(ii) Image ∂_{q+1}: $\bar{G}_{q+1} \to \bar{G}_q \subset \operatorname{Ker} \partial_q$: $\bar{G}_q \to \bar{G}_{q-1}$, $q > 0$.

(iii) $\partial_{q+1}(\bar{G}_{q+1})$ is a normal subgroup of $\bar{G}_q$ and of G_q.

Proof: Let $x \in \bar{G}_{q+1}$. Then $\partial_i \partial_{q+1} x = \partial_q \partial_i = e_q$, $0 \leq i \leq q$, and this proves (i) and (ii). Let $z \in G_q$ and define $w = s_q z \cdot x \cdot s_q z^{-1}$. $\partial_i w = e_q$ for $i \leq q$, hence $w \in \bar{G}_{q+1}$, and $\partial_{q+1} w = z \cdot \partial_{q+1} x \cdot z^{-1}$. This proves (iii).

Now $\bar{G}$ is a (not necessarily Abelian) chain complex with respect to the last face operator. Define $\pi_q'(G) = H_q(\bar{G})$. Then we find

PROPOSITION 17.4: $\pi_q(G) = \pi_q'(G)$ for all q.

Proof: $\tilde{G}_q = Z_q \bar{G}$, hence if $x \in \tilde{G}_q$, x represents an element of $\pi_q'(G)$. If $x, y \in \tilde{G}_q$ and z: $x \sim y$, then $s_q x^{-1} \cdot z \in \bar{G}_{q+1}$ and $\partial_{q+1}(s_q x^{-1} \cdot z) = x^{-1} y$, so that x and y represent the same element of $\pi_q'(G)$. Using Proposition 17.2, it follows that there is a natural epimirphism of groups i: $\pi_q(G) \to \pi_q'(G)$. Suppose $i[x] = 0$. Then there exists $z \in \bar{G}_{q+1}$ such that $\partial_{q+1} z = x$, hence z: $e_q \sim x$ and $[x] = 0$ in $\pi_q(G)$. This completes the proof.

PROPOSITION 17.5: A group complex G is minimal if and only if ∂_{q+1}: $\bar{G}_{q+1} \to \bar{G}_q$ is zero for all q.

Proof: Suppose G is minimal and $x \in \bar{G}_{q+1}$. Then

$$\partial_i x = e_q = \partial_i e_{q+1}, \quad i \leq q,$$

and therefore $\partial_{q+1} x = \partial_{q+1} e_{q+1} = e_q$. Conversely, suppose ∂_{q+1}: $\bar{G}_{q+1} \to \bar{G}_q$ is zero for all q and let $x, y \in G_{q+1}$ satisfy $\partial_i x = \partial_i y$, $i \neq k$. Oberve that $\partial_i(xy^{-1}) = e_q$, $i \neq k$. If $k = q+1$, let $z = xy^{-1}$. If $k \leq q$ and $q - k$ is even, let

$$z = (s_q \partial_k xy^{-1})(s_{q-1} \partial_k yx^{-1}) \cdots (s_k \partial_k xy^{-1})(yx^{-1}).$$

If $k \le q$ and $q - k$ is odd, let

$$z = (s_q \partial_k xy^{-1})(s_{q-1} \partial_k yx^{-1}) \cdots (s_k \partial_k yx^{-1})(xy^{-1}) \ .$$

In all cases, $\partial_i z = e_q$, $i < q+1$, and $\partial_{q+1} z = \partial_k(xy^{-1})$. By assumption, $\partial_{q+1} z = e_q$, so that $\partial_k x = \partial_k y$, as was to be proven.

The result above states that a group complex G is minimal if and only if the corresponding chain complex $\overline{G}$ is minimal.

§18. Principal fibrations and twisted Cartesian products.

We will first define principal fibrations and then twisted Cartesian products. It will turn out that the former concept is a special case of the latter.

DEFINITIONS 18.1: A group complex G is said to operate from the left on a complex E if there exists a simplicial map ϕ: $G \times E \to E$ such that $\phi(e_q, x) = x$ and

$$\phi(g_1, \phi(g_2, x)) = \phi(g_1 g_2, x) \ .$$

We will denote $\phi(g, x)$ by gx. G is said to operate effectively if $gx = x$ for all $x \in E_q$ implies $g = e_q$. G is said to operate principally if $gx = x$ for any one $x \in E_q$ implies $g = e_q$. Note that G operates principally on itself from both the left and the right. If G operates principally from the right on E, then we define a quotient complex B of E by identifying x and xg for all $x \in E_q$ and $g \in G_q$. The map p: $E \to B$ is called a principal fibration with base B and structural group G.

LEMMA 18.2: Any principal fibration is a Kan fibration.

Proof: Let p: $E \to B$ be a principal fibration with group G. Suppose $x_0, \ldots, x_{k-1}, x_{k+1}, \ldots, x_{q+1} \in E_q$ satisfy

$$\partial_i x_j = \partial_{j-1} x_i, \quad i < j, \quad i, j \neq k,$$

and $b \in B_{q+1}$ satisfies $\partial_i b = p(x_i)$, $i \neq k$. Then if $p(y) = b$, $y \in E_{q+1}$, $\partial_i y = x_i g_i$ for some $g_i \in G_q$, $i \neq k$. If $i < j$, $i, j \neq k$, we have

$$\partial_i (x_j g_j) = \partial_i \partial_j y = \partial_{j-1} \partial_i y = \partial_{j-1} (x_i g_i) .$$

Since $\partial_i x_j = \partial_{j-1} x_i$ and G acts principally, $\partial_i g_j = \partial_{j-1} g_i$ and therefore there exists $g \in G_{q+1}$ such that $\partial_i g = g_i$, $i \neq k$. Let $x = yg^{-1}$. Then $\partial_i x = x_i$, $i \neq k$, and $p(x) = b$, as desired.

DEFINITIONS 18.3: Let F and B be complexes and G a group complex which operates on F from the left. A twisted Cartesian product, or $T\ C\ P$, with fibre F, base B, and group G is a complex, denoted by either $F \times_\tau B$ or $E(\tau)$, which satisfies

$$E(\tau)_n = F_n \times B_n$$

and has face and degeneracy operators

(i) $\partial_i (f, b) = (\partial_i f, \partial_i b), \quad i > 0;$

(ii) $\partial_0 (f, b) = (\tau(b) \cdot \partial_0 f, \partial_0 b), \quad \tau(b) \in G;$

(iii) $s_i (f, b) = (s_i f, s_i b), \quad i \geq 0;$

$\tau\colon B_q \to G_{q-1}$ is called the twisting function. The requirement that $E(\tau)$ be a complex is equivalent to the following identities on τ:

$$\text{(T)} \qquad \begin{aligned} \partial_0 \tau(b) &= [\tau(\partial_0 b)]^{-1} \tau(\partial_1 b) \\ \partial_i \tau(b) &= \tau(\partial_{i+1} b), \quad i > 0 \\ s_i \tau(b) &= \tau(s_{i+1} b), \quad i \geq 0 \\ \tau(s_0 b) &= e_q, \quad b \in B_q . \end{aligned}$$

If $F = G$, then $E(\tau)$ is called a principal $T\ C\ P$, or $PT\ C\ P$. The term $T\ C\ P$ will also stand for the projection $p\colon E(\tau) \to B$.

PROPOSITION 18.4: Let $p\colon E(\tau) \to B$ be a $T\ C\ P$ with group G and suppose the fibre F is a Kan complex. Then:

(i) p is a Kan fibration;

(ii) p is minimal if and only if F is a minimal complex;

(iii) if $F = G$, p is a principal fibration.

Proof: (i) Suppose $x_0, \ldots, x_{k-1}, x_{k+1}, \ldots, x_{q+1} \in E(\tau)_q$ satisfy $\partial_i x_i = \partial_{j-1} x_i$, $i < j$, $i, j \neq k$, and $b \in B_{q+1}$ satisfies $\partial_i b = p(x_i)$, $i \neq k$. If $x_i = (f_i, b_i)$, we have $\partial_i b = b_i$, $i \neq k$. If $k = 0$, then the f_i are clearly compatible, and we may choose $f \in F_{q+1}$ such that $\partial_i f = f_i$, $i \neq 0$. If $k > 0$, then it is easily verified that $[\tau(b)^{-1}] \cdot f_0$ and the f_i, $i > 0$, $i \neq k$, are compatible, and we may choose $f \in F_{q+1}$ such that $\partial_0 f = [\tau(b)^{-1}] \cdot f_0$ and $\partial_i f = f_i$, $i > 0$, $i \neq k$. In both cases $x = (f, b)$ satisfies $\partial_i x = x_i$, $i \neq k$, and $p(x) = b$, as desired.

(ii) If ϕ is any vertex of B, then $f \to (f, \phi)$ defines an isomorphism of F with the fibre over ϕ. Therefore if p is minimal, F is a minimal complex by Lemma 10.2. Conversely, suppose F is minimal and let $x = (f, b)$ and $x' = (f', b')$ satisfy $\partial_i x = \partial_i x'$, $i \neq k$, and $p(x) = p(x')$. Then $b = b'$ and $\partial_i f = \partial_i f'$ if $i \neq k$, hence $\partial_k f = \partial_k f'$. It follows that

$$\partial_k x = \partial_k x',$$

as desired.

(iii) Since $F = G$, G operates on the right of $E(\tau)$ via $(g_1, b)g_2 = (g_1 g_2, b)$. Clearly B may be identified with the quotient complex of $E(\tau)$ obtained by identifying x and xg, and the result follows.

Now we wish to obtain a converse to (iii) of the proposition. We need a preliminary result.

DEFINITION 18.5: Let $p\colon E \to B$ be a simplicial map. A pseudo-cross section of p is a function $\sigma\colon B \to E$ satisfying $p\sigma = 1_B$, $\partial_i\sigma = \sigma\partial_i$ if $i > 0$, and $s_i\sigma = \sigma s_i$ if $i \geq 0$. If $\partial_0\sigma = \sigma\partial_0$, then σ is called a cross section. Observe that any $PTCP$ has the canonical pseudo-cross section σ defined by $\sigma(b) = (e_p, b)$, $b \in B_q$, and that σ and the twisting function τ are related by $\partial_0\sigma(b) = \sigma(\partial_0 b)\cdot\tau(b)$.

LEMMA 18.6: Let $p\colon E \to B$ be a Kan fibration, where p is onto. Let $B' \subset B$ and suppose $\sigma'\colon B' \to p^{-1}(B')$ is a pseudo-cross section (B' may be empty). Then there exists at least one pseudo-cross section $\sigma\colon B \to E$ such that $\sigma|B' = \sigma'$.

Proof: If $b \in B_0$, let $\sigma(b)$ be any element $x \in E_0$ such that $p(x) = b$, choosing $\sigma'(b)$ if $b \in B_0'$. Now suppose σ has been defined on each B_q, $q < n$, and $n > 0$. Let $b \in B_n$. If $b \in B_n'$, define $\sigma(b) = \sigma'(b)$. Assume $b \notin B_n'$. If b is degenerate, say $b = s_i y$, define $\sigma(b) = s_i\sigma(y)$. If $b = s_j z$ is also true, then it is easily verified that $s_i\sigma(y) = s_j\sigma(z)$. Finally, suppose b is non-degenerate. There exists $x \in E_n$ such that $p(x) = b$ and $\partial_i x = \sigma(\partial_i b)$ for $i > 0$. Let $\sigma(b) = x$. Clearly $p\sigma(b) = x$ and $\partial_i\sigma(b) = \sigma(\partial_i b)$, $i > 0$, and this procedure does define a pseudo-cross section.

Together with the preceding lemma and Lemma 18.2, the following proposition implies that every principal fibration is a $PTCP$.

PROPOSITION 18.7: A principal fibration $p\colon E \to B$ with group G and pseudo-cross section σ may be identified with the $PTCP$ $p\colon E(\tau) \to B$ with group G and twisting function determined by the formula $\partial_0\sigma(b) = \sigma(\partial_0 b)\tau(b)$.

Proof: We note first that $p\sigma(\partial_0 b) = \partial_0 b = p(\partial_0\sigma(b))$

implies $\partial_0 \sigma(b) = \sigma(\partial_0 b)\tau(b)$ for some $\tau(b) \in G$. Now define a bijection of graded sets $\psi\colon G \times B \to E$ by $\psi(g, b) = \sigma(b)g$ and give $G \times B$ the structure of complex induced by ψ^{-1}. Then it is easily verified that $\psi^{-1}(E) = G \times_\tau B$, as desired. Note that $\psi\sigma = \sigma$, where σ on the left is the canonical pseudo-cross section of $E(\tau)$.

REMARKS 18.8: Clearly a principal fibration (or fibre bundle) in the category of topological spaces passes via the functor S to a principal fibration in the category of simplicial sets. This gives one motivation for the study of $PTCP$'s.

§19. The group of a fibre bundle.

In this section, we show how to associate a group to a fibre bundle. The theory here is roughly analogous to that of coordinate transformations in the topological case. As an easy consequence of this theory, we shall find that every fibre bundle may be regarded as a TCP.

We need some preliminary observations about the operations of groups on complexes. In Lemma 6.12, we showed that K^K is a monoid complex if K is a complex. The product is defined by $(fg)(x, u) = f(g(x, u), u)$, where $f, g \in (K^K)_n$, $x \in K_q$, and $u \in \Delta[n]_q$. If E is another complex and $f \in E^K$, then the same formula defines an operation from the right of K^K on E^K. If $g \in (K^K)_n$, then G determines $\tilde{g}\colon K \times \Delta[n] \to K \times \Delta[n]$, where $\tilde{g}(x, u) = (g(x, u), u)$; $\tilde{g}$ is an isomorphism if and only if g is invertible. We let $A(K) \subset K^k$ denote the set of invertible elements. $A(K)$ is a group complex. Observe that K^K and $A(K)$ operate from the left on K via $f \cdot x = f(x, \Delta_n)$, $x \in K_n$, $f \in (K^K)_n$. If a group complex G operates on K, we define a homomorphism $\nu\colon G \to A(K)$ by

$v(g)(x, u) = (\varepsilon g) \cdot x$, where $g \in G_n$, $x \in K_q$, $u \in \Delta[n]_q$, and ε is that simplicial operator such that $u = \varepsilon \Delta_n$. Clearly

$$g \cdot x = v(g) \cdot x .$$

v is a monomorphism if and only if G operates effectively. Replacing G by $G/\mathrm{Ker}\, v$, we may assume that operations by groups are always effective.

Now let $p\colon E \to B$ be a fibre bundle with fibre F (Definition 11.8). For each $b \in B_n$, choose a strong isomorphism

$$a(b)\colon F \times \Delta[n] \to E^{\bar{b}} .$$

$\{a(b)\}$ is called an atlas of p. If for each $b \in B_n$ we are given $g(b) \in A(F)_n$, then $\bar{a}(b) = a(b)\tilde{g}(b)$ defines another atlas $\{\bar{a}(b)\}$. Conversely, if $\{a(b)\}$ and $\{\bar{a}(b)\}$ are atlases of p, then letting $\tilde{g}(b) = a(b)^{-1}\bar{a}(b)$, we have $\bar{a}(b) = a(b)\tilde{g}(b)$.

Given an atlas $\{a(b)\}$, let $\beta(b) = \tilde{b} \circ a(b)\colon F \times \Delta[n] \to E$, where $\tilde{b}\colon E^{\bar{b}} \to E$ covers $\bar{b}\colon \Delta[n] \to B$, $b \in B_n$. Since

$$a(b)(f, u) = (\beta(b)(f, u), u) ,$$

$\{a(b)\}$ and $\{\beta(b)\}$ determine each other. We regard $\{\beta(b)\}$ to be contained in E^F, and then face and degeneracy operators and operations from the right by elements of $A(F)$ are defined on elements of $\{\beta(b)\}$, although $\{\beta(b)\}$ need not be closed under these operations. If $\{\bar{a}(b\}$ and $\{a(b)\}$ are atlases related by $\bar{a}(b) = a(b)\tilde{g}(b)$, then $\bar{\beta}(b) = \beta(b)g(b)$.

The atlas $\{a(b)\}$ is said to be normalized if

$$\beta(s_i b) = s_i \beta(b)$$

for all i and b. If we choose an $a(b)$ for each non-degenerate b and define $\beta(s_i b) = s_i \beta(b)$ for all i and b, we obtain a normalized atlas. The definition is consistent since if $s_i b = s_j b'$, $i < j$, then, using induction on the dimension of b,

$$s_j\beta(b') = s_j\beta(s_i\partial_j b) = s_j s_i\beta(\partial_j b) = s_i\beta(s_{j-1}\partial_j b) = s_i\beta(b).$$

From now on, we assume all atlases to be normalized.

The difference between $\partial_i\beta(b)$ and $\beta(\partial_i b)$ is more significant, and we investigate this now. Recall the definition (page 14) of $\delta_i\colon \Delta[n-1] \to \Delta[n]$. If $b \in B_n$, then since

$$a(b)(1\times\delta_i)\colon\ F\times\Delta[n-1] \to E^{\bar b}$$

is an isomorphism onto $\text{Image}(1\times\delta_i)$, $1\times\delta_i\colon\ E^{\overline{\partial_i b}} \to E^{\bar b}$, there exists a strong isomorphism $a_i(b)\colon\ F\times\Delta[n-1] \to E^{\overline{\partial_i b}}$ such that the following diagram commutes:

$$\begin{array}{ccccc} F\times\Delta[n] & \xrightarrow{a(b)} & E^{\bar b} & \xrightarrow{\tilde b} & E \\ \uparrow{\scriptstyle 1\times\delta_i} & & \uparrow{\scriptstyle 1\times\delta_i} & \nearrow{\scriptstyle \widetilde{\partial_i b}} & \\ F\times\Delta[n-1] & \xrightarrow{a_i(b)} & E^{\overline{\partial_i b}} & & \end{array}$$

By definition, $\widetilde{\partial_i b}\circ a_i(b) = \partial_i\beta(b)$. Define $\tau_i(b) \in A(F)_{n-1}$ by $\widetilde{\tau_i(b)} = [a(\partial_i b)]^{-1}a_i(b)\colon\ F\times\Delta[n-1] \to F\times\Delta[n-1]$. Since $a_i(b) = a(\partial_i b)\widetilde{\tau_i(b)}$, $\partial_i\beta(b) = \beta(\partial_i b)\tau_i(b)$. $\{\tau_i(b)\}$ is called the set of transformation elements associated with the atlas $\{a(b)\}$.

DEFINITIONS 19.1: Let $p\colon\ E \to B$ be a fibre bundle with fibre F and let G be a subgroup complex of $A(F)$. An atlas $\{a(b)\}$ of p is called a G-atlas if all of its transformation elements are in G. Two G-atlases $\{a(b)\}$ and $\{\bar a(b)\}$ are said to be G-equivalent if for all $b \in B$, $\bar a(b) = a(b)\tilde g(b)$, where $g(b) \in G$. $p\colon\ E \to B$ together with a G-equivalence class of G-atlases is called a G-bundle. Observe that p is necessarily an $A(F)$-bundle. Let $p\colon\ E \to B$ and $p'\colon\ E' \to B'$ be G-bundles with the same fibre F. A map $(\bar f, f)\colon\ (E, p, B) \to (E', p', B')$ is said to be a map of

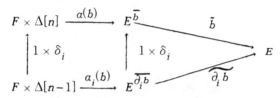

G-bundles if for all $b \epsilon B$ and any G-atlases $\{\alpha(b)\}$ and $\{\alpha'(b')\}$ in the given G-equivalence classes of atlases,

$$\tilde{f} \circ \beta(b) = \beta'(f(b))\psi(b),$$

where $\psi(b) \epsilon G$. There results a concept of G-equivalence of G-bundles. The concept of $A(F)$-equivalence is identical with that of isomorphism of fibre bundles with a given fibre.

An atlas $\{\alpha(b)\}$ of a fibre bundle $p: E \to B$ is said to be regular if $\tau_i(b) = e_{n-1}$ for all $i > 0$ and $b \epsilon B_n$, $n \geq 1$; this means that $\partial_i \beta(b) = \beta(\partial_i b)$, $i > 0$.

LEMMA 19.2: Let $p: E \to B$ be a fibre bundle with fibre F and let G be a subgroup complex of $A(F)$. Then there is a regular G-atlas in every G-equivalence class of G-atlases of p.

Proof: Let $\{\alpha(b)\}$ be a G-atlas of p. We will define a G-equivalent regular G-atlas $\{\bar{\alpha}(b)\}$. If $b \epsilon B_0$, let $\bar{\alpha}(b) = \alpha(b)$. Suppose $\bar{\alpha}(b)$ has been defined for $b \epsilon B_{n-1}$, $n \geq 1$. Let $b \epsilon B_n$. If b is degenerate, $\bar{\alpha}(b)$ is already defined and has the desired property. Suppose b is non-degenerate and

$$\bar{\alpha}(\partial_i b) = \alpha(\partial_i b)\tilde{g}_i, \qquad i > 0.$$

If $0 < i < j$, then an application of the induction hypothesis gives

$$\begin{aligned}(1) \quad \partial_i \partial_j \beta(b) &= \partial_i \beta(\partial_j b) \cdot \partial_i \tau_j(b) = \partial_i \bar{\beta}(\partial_j b) \cdot (\partial_i g_j)^{-1} \cdot \partial_i \tau_j(b) \\ &= \partial_{j-1} \bar{\beta}(\partial_i b) \cdot (\partial_i g_j)^{-1} \cdot \partial_i \tau_j(b) \\ &= \partial_{j-1} \beta(\partial_i b)(\partial_{j-1} g_i)(\partial_i g_j)^{-1} \cdot \partial_i \tau_j(b) \\ &= \partial_{j-1} \partial_i \beta(b)[\partial_{j-1}\tau_i(b)]^{-1}(\partial_{j-1} g_i)(\partial_i g_j)^{-1} \partial_i \tau_j(b).\end{aligned}$$

Clearly G operates principally on the G-submodule of E^F generated by $\{\beta(b)\}$ and therefore $\partial_i(g_j^{-1} \tau_j(b)) = \partial_{j-1}(g_i^{-1}\tau_i(b))$. Choose $g \epsilon G_n$ such that $\partial_i g = g_i^{-1} \tau_i(b)$ for $i > 0$ and let $\bar{\alpha}(b) = \alpha(b)\tilde{g}^{-1}$. Then:

(2) $\partial_i \bar{\beta}(b) = \partial_i \beta(b)(\partial_i g)^{-1} = \beta(\partial_i b)\tau_i(b)[\tau_i(b)]^{-1} g_i = \bar{\beta}(\partial_i b)$.

This completes the proof.

If $G' \subset G$ is a sub-group complex and there exists a G'-atlas in the given G-equivalence class of G-atlases of a G-bundle, then we say that the group of the bundle can be reduced to G'. Note that the resulting G'-bundle need not be uniquely determined.

LEMMA 19.3: Let $p: E \to B$ be a G-bundle. Suppose $G' \subset G$ is a subgroup complex which is a deformation retract of G. Then the group of p can be reduced to G'.

Proof: Let $\{a(b)\}$ be a regular G-atlas in the given G-equivalence class of G-atlases of p. We will define a G-equivalent regular G'-atlas $\{\bar{a}(b)\}$. Let $\bar{a}(b) = a(b)$ if $b \in B_0$ and suppose $\bar{a}(b)$ has been defined for $b \in B_{n-1}$, $n \geq 1$. Let $b \in B_n$. If b is degenerate, $\bar{a}(b)$ is already defined and has the desired properties. Suppose b is nondegenerate and $\bar{a}(\partial_i b) = a(\partial_i b)\tilde{g}_i$, $i \geq 0$. Arguing as in (1) of the previous lemma, we find that if $0 < i < j$, then

(1) $$\partial_i \partial_j \beta(b) = \partial_{j-1} \partial_i \beta(b)(\partial_{j-1} g_i)(\partial_i g_j)^{-1} .$$

Similarly we find, for $0 < j \leq n$,

(2) $$\partial_0 \partial_j \beta(b) = \partial_{j-1}\partial_0 \beta(b)[\partial_{j-1} \tau_0(b)]^{-1} \partial_{j-1} g_0[\bar{\tau}_{j-1}(\partial_0 b)]^{-1} \bar{\tau}_0(\partial_j b)(\partial_0 g_j)^{-1}.$$

It follows that $\partial_i g_j = \partial_{j-1} g_i$, $0 < i < j$, and

(3) $$\bar{\tau}_{j-1}(\partial_0 b)^{-1} \bar{\tau}_0(\partial_j b) = (\partial_{j-1} g_0)^{-1} \partial_{j-1} \tau_0(b) \cdot \partial_0 g_j, \quad 0 < j \leq n.$$

Choose $g \in G_n$ such that $\partial_i g = g_i$, $i > 0$, and let $a'(b) = a(b)\tilde{g}$. Then $\partial_i \beta'(b) = \bar{\beta}(\partial_i b)$, $i > 0$, and

$$\partial_0 \beta'(b) = \bar{\beta}(\partial_0 b) g_0^{-1} \tau_0(b) \partial_0 g .$$

Let $u = g_0^{-1}\tau_0(b)\partial_0 g$ and observe that by (3) and by $\partial_0 g_j = \partial_{j-1}\partial_0 g$, we have $\partial_{j-1} u \in G'_{n-1}$, $0 < j \leq n$. Now let $h\colon G \to G$ be a homotopy such that $h(G') \subset G'$ and $\partial_0 h_0(x) = x$ and $\partial_{q+1} h_q(x) \in G'_q$ if $x \in G_q$. If n is odd, let $k = h_0(u)^{-1} h_1(u) h_2(u)^{-1} \cdots h_{n-1}(u)^{-1}$ and if n is even, let $k = h_0(u)^{-1} h_1(u) h_2(u)^{-1} \cdots h_{n-1}(u)$. Define $\alpha''(b) = \alpha'(b)\tilde{k}$. Then $\partial_i \beta''(b) = \bar{\beta}(\partial_i b)\partial_i k$ if $i > 0$ and

$$\partial_0 \beta''(b) = \bar{\beta}(\partial_0 b)\cdot u \cdot \partial_0 k \ .$$

By the identities satisfied by h, it follows that

$$\partial_i \beta''(b) = \bar{\beta}(\partial_i b) g_i', \quad g_i' \in G'_{n-1}, \quad i \geq 0 \ .$$

Thus $\tau_i''(b) \in G'_{n-1}$ for all i and, arguing as in the proof of the previous lemma, we may replace $\beta''(b)$ by $\bar{\beta}(b)$ such that $\bar{\tau}_i(b) = e_{n-1}$ for $i > 0$ and $\bar{\tau}_0(b) \in G'_{n-1}$.

From now on, we assume all atlases to be regular and we let $\tau(b)$ denote the transformation elements $\tau_0(b)$ of an atlas $\{\alpha(b)\}$ of a G-bundle $p\colon E \to B$. We call τ a twisting function of p. This notation is justified by the following result.

THEOREM 19.4: Let $p\colon E \to B$ be a G-bundle with fibre F and let $\{\alpha(b)\}$ be a G-atlas of p. Then the transformation elements $\tau(b)$ define a twisting function τ and (E, p, B) is isomorphic to $(F \times_\tau B, p, B)$.

Proof: Define $\xi\colon F_n \times B_n \to E_n$ by $\xi(f, b) = \beta(b)(f, \Delta_n)$. ξ is clearly an isomorphism of sets and:

$$\partial_0 \xi(f, b) = \xi(\tau(b)\partial_0 f, \partial_0 b) \ ,$$
$$\partial_i \ \xi(f, b) = \xi(\partial_i f, \partial_i b) \ , \quad i > 0 \ ,$$
$$s_i \ \xi(f, b) = \xi(s_i f, s_i b) \ , \quad i \geq 0 \ .$$

Thus if ξ^{-1} is used to define a structure of complex on the set

$F \times B$, we obtain $\xi^{-1}(E) = F \times_\tau B$. This proves the result.

Observe that the G-bundle $p\colon E \to B$ with atlas $\{\alpha(b)\}$ determines the isomorphism ξ.

§20. Fibre bundles and twisted Cartesian products

In this section we first define maps of TCP's with fixed fibre and group. We then examine in detail the relationship between TCP's and fibre bundles and prove that every TCP is a fibre bundle.

DEFINITIONS 20.1: Let $p\colon E(\tau) \to B$ and $p'\colon E(\tau') \to B'$ be TCP's with group G and fibre F. A simplicial map

$$\theta\colon E(\tau) \to E(\tau')$$

is said to be a map of TCP's if $\theta(f, b) = (\psi(b)f, \pi(b))$, where $\psi\colon B \to G$ is a map of sets. Observe that $p'\theta = \pi p$. The requirement that θ be a simplicial map is equivalent to the following identities on ψ:

$$\text{(U)}\qquad \begin{aligned} \tau'\pi(b)\partial_0\psi(b) &= \psi(\partial_0 b)\tau(b) \\ \partial_i\psi(b) &= \psi(\partial_i b) \quad \text{if } i > 0 \\ s_i\psi(b) &= \psi(s_i b) \quad \text{if } i \geq 0. \end{aligned}$$

θ is said to be a strong map if $B = B'$ and $\pi = 1_B$, and in this case we say that τ and τ' are equivalent twisting functions. Observe that every strong map is necessarily a strong isomorphism.

LEMMA 20.2: Let $p\colon E \to B$ and $p'\colon E' \to B$ be G-bundles with fibre F. Then p and p' are strongly G-equivalent if and only if τ and τ' are equivalent, where τ and τ' are any twisting functions of p and p'.

Proof: Suppose p and p' are strongly G-equivalent. Then there exists a simplicial isomorphism $\Lambda\colon E \to E'$ such that

$p'\Lambda = p$ and $\Lambda\beta(b) = \beta'(b)\psi(b)$, $\psi(b) \in G$, where $\{a(b)\}$ and $\{a'(b)\}$ are any G-atlases of p and p'. If τ and τ' are the twisting functions defined by $\{a(b)\}$ and $\{a'(b)\}$, then it is easily verified that ψ satisfies the identities (U) (with $\pi = 1_B$). Thus τ and τ' are equivalent. Conversely, suppose τ and τ' are equivalent and $\psi: B \to G$ satisfies the identities (U). Then if $x \in E_n$ is written as $x = \beta(b)(f, \Delta_n)$, $b = p(x)$, and we define

$$\Lambda(x) = \beta'(b)\psi(b)(f, \Delta_n) = \beta'(b)(\psi(b)f, \Delta_n),$$

it follows that Λ is a strong G-equivalence of G-bundles.

Before proving that every TCP is a fibre bundle, we need the following definitions.

DEFINITIONS 20.3: Let G operate on the right of a complex E and on the left of a complex F. Then $F \times_G E$ is defined to be the quotient of $F \times E$ obtained by identifying (gf, x) and (f, xg) for all x, f, and g. If $E = G \times_\tau B$, then $F \times_G E$ may be identified with $F \times_\tau B$ and is called the TCP with fibre F associated to $G \times_\tau B$. If $p: E \to B$ is a principal G-bundle, then for some τ E is isomorphic to $G \times_\tau B$, and therefore G operates on the right of E. (The operation is actually independent of the choice of atlas.) In this case, $(F \times_G E, p^*, B)$, where $p^*(f, x) = p(x)$, is said to be the G-bundle with fibre F associated to p. If $\{a(b)\}$ is an atlas of p, then $\{a^*(b)\}$ is an atlas of p^*, where

$$\beta^*(b)(f, u) = (f, \beta(b)(e_q, u)), \quad u \in \Delta[n]_q, \quad b \in B_n, \quad f \in F_q.$$

Now observe that if $F \times_\tau B$ is a TCP and $\pi: A \to B$ is a simplicial map, then $E(\tau)^\pi$ may be identified with $F \times_{\tau^\pi} A$, where $\tau^\pi = \tau \circ \pi$, and $\tilde{\pi}$: $E(\tau)^\pi \to E(\tau)$ is a map of TCP's.

THEOREM 20.4: Let $F \times_\tau B$ be a TCP and suppose $\lambda \simeq \pi: A \to B$. Then $E(\tau)^\lambda$ is strongly isomorphic to $E(\tau)^\pi$.

Proof: Suppose first that $F = G$, that is, $E(\tau)$ is a PTCP. Then by Proposition 11.5, there exists a strong homotopy equivalence $\theta: E(\tau)^\lambda \to E(\tau)^\pi$. Clearly we may write $\theta(g, a) = (\psi(a)g, a)$. Thus θ is a strong map of TCP's, hence a strong isomorphism, and the twisting functions τ^λ and τ^π are equivalent. The latter statement implies the result for arbitrary F, and the theorem is proven.

COROLLARY 20.5: Every TCP is a fibre bundle.

Proof: Let $F \times_\tau B$ be a TCP. Choose a base point $\phi \in B_0$, and let $b \in B_n$. $\bar{b} \simeq \bar{\phi}: \Delta[n] \to B$, hence $E(\tau)^{\bar{b}}$ is strongly isomorphic to $E(\tau)^{\bar{\phi}}$. Since $\tau^{\bar{\phi}} = \tau \circ \phi$, the latter is $F \times \Delta[n]$. This proves the result.

REMARKS 20.6: An atlas for $F \times_\tau B$ may of course be defined by $\beta(b)(f, \Delta_n) = (f, b)$. The twisting function of this atlas is τ. In other words, a TCP with group G is essentially just a G-bundle with a chosen atlas.

If $p: E \to B$ is a G-bundle with given atlas $\{\alpha(b)\}$ and $\pi: A \to B$ is a simplicial map, then $p^\pi: E^\pi \to A$ is a G-bundle with atlas $\{\alpha^\pi(a)\}$, where $\tilde{\pi}\beta^\pi(a) = \beta(\pi(a))$. If p is principal, then so is p^π and $\tilde{\pi}$ is equivariant, that is, $\tilde{\pi}(xg) = \tilde{\pi}(x)g$ for $x \in E^\pi$, $g \in G$.

COROLLARY 20.6: If $p: E \to B$ is a G-bundle and $\lambda \simeq \pi: A \to B$, then $p^\lambda: E^\lambda \to A$ and $p^\pi: E^\pi \to A$ are strongly G-equivalent G-bundles.

Finally, we note that since every TCP is associated to a PTCP and every TCP is a fibre bundle, we have

PROPOSITION 20.7: Every G-bundle is associated to a principal G-bundle which is uniquely determined up to strong G-equivalence of G-bundles.

The converse is obvious: If $p: E \to B$ is a principal G-bundle and G operates on F from the left, then there is a unique G-bundle with fibre F associated to p.

§21. Universal bundles and classifying complexes

We will construct a PTCP, or G-bundle,

$$W(G) = G \times_{\tau(G)} \overline{W}(G)$$

for each simplicial group G. $\overline{W}(G)$ will play the role of a classifying complex for G.

DEFINITION 21.1: A PTCP $G \times_\tau B$ is said to be of type (W) if B_0 has one element b_0 and if ∂_0: $e_q \times B \to E(\tau)_{q-1}$ is an isomorphism of sets for all $q \geq 1$. We let S: $E(\tau)_{q-1} \to e_q \times B$ denote the isomorphism of sets inverse to ∂_0, so that $\partial_0 S$ is the identity on $E(\tau)_{q-1}$.

We will prove that, up to natural isomorphism, there exists one and only one PTCP of type (W) with group G. Before proving existence, we obtain some of the properties of PTCP's of type (W).

LEMMA 21.2: If $G \times_\tau B$ is a PTCP of type (W), then:

(i) $\partial_1 S(x) = (e_0, b_0)$ if $x \in E(\tau)_0$;

(ii) $\partial_{i+1} S(x) = S(\partial_i x)$ if $x \in E(\tau)_q$, $q > 0$;

(iii) $S = s_0$ on $e_q \times B_q$;

(iv) $s_{i+1} S(x) = S(s_i x)$ if $x \in E(\tau)_q$, $q \geq 0$.

Proof: If $x \in E(\tau)_q$, $q \geq 0$, then $x = (\tau(b), \partial_0 b)$ for some $b \in B_{q+1}$. If $i > 0$,

$$S(\partial_i x) = S(\partial_i \tau(b), \partial_i \partial_0 b) = S(\tau(\partial_{i+1} b), \partial_0 \partial_{i+1} b) = (e_q, \partial_{i+1} b),$$

and

$$S(\partial_0 x) = S(\tau(\partial_0 b) \partial_0 \tau(b), \partial_0 \partial_0 b) = S(\tau_1(b), \partial_0 \partial_1 b) = (e_q, \partial_1 b) .$$

On the other hand, $\partial_i S(x) = \partial_i(e_{q+1}, b) = (e_q, \partial_i b)$ if $i > 0$. This proves (i) and (ii), and the proof of (iv) is similar. (iii) follows immediately from the definition.

LEMMA 21.3: If $G \times_\tau B$ is a PTCP of type (*W*), then B, and therefore $E(\tau)$, is a Kan complex.

Proof: Let $b_0, \ldots, b_{k-1}, b_{k+1}, \ldots, b_{q+1} \in B_q$ satisfy

$$\partial_i b_j = \partial_{j-1} b_i, \quad i < j, \quad i, j \neq k.$$

We must find b such that $\partial_i b = b_i$, $i \neq k$. We break the proof into two cases.

(i) $k > 0$: Here we let $x_i = (\tau(b_{i+1}), \partial_0 b_{i+1})$. Then $\partial_i x_j = \partial_{j-1} x_i$, $i < j$, $i, j \neq k-1$, and $p(x_i) = \partial_0 b_{i+1} = \partial_i b_0$. Since p is a Kan fibration, there exists $x \in E(\tau)_q$ such that $\partial_i x = x_i$, $i \neq k-1$, and $p(x) = b_0$. Let $(e_{n+1}, b) = S(x)$. Then $\partial_0 b = p \partial_0 S(x) = p(x) = b_0$ and, if $i > 0$, $i \neq k$, we have

$$\partial_i b = p \partial_i S(x) = p S(\partial_{i-1} x) = p(e_n, b_i) = b_i .$$

(ii) $k = 0$: Here we note that $\partial_i \tau(b_{j+1}) = \partial_{j-1} \tau(b_{i+1})$ if $0 < i < j$, so that there exists $g \in G_q$ such that

$$\partial_i g = \tau(b_{i+1}), \quad i > 0 .$$

Let $(e_{q+1}, b) = S(g, c)$, where $c = p S(\tau(b_1)(\partial_0 g)^{-1}, \partial_0 b_1)$. Now $\tau(c) = \tau(b_1)(\partial_0 g)^{-1}$ and $\partial_0 c = \partial_0 b_1$. Further:

$$\begin{aligned} \partial_1 c &= p S(\tau(\partial_0 b_1) \partial_0 \tau(b_1)(\partial_0 \partial_0 g)^{-1}, \partial_0 \partial_0 b_1) \\ &= p S(\tau(\partial_1 b)[\partial_0 \tau(b_2)]^{-1}, \partial_0 \partial_1 b_1) \\ &= p S(\tau(\partial_0 b_2), \partial_0 \partial_0 b_2) = \partial_0 b_2 , \end{aligned}$$

and

$$\begin{aligned}\partial_i c &= pS(\partial_{i-1}\tau(b_1)(\partial_{i-1}\partial_0 g)^{-1}, \partial_{i-1}\partial_0 b_1)\\ &= pS(\tau(\partial_i b_1)[\partial_0 \tau(b_{i+1})]^{-1}, \partial_0\partial_i b_1)\\ &= pS(\tau(\partial_0 b_{i+1}), \partial_0\partial_0 b_{i+1}) = \partial_0 b_{i+1}, \quad i > 1.\end{aligned}$$

Therefore $\partial_1 b = pS(\tau(c)\partial_0 g, \partial_0 c) = pS(\tau(b_1), \partial_0 b_1) = b_1$, and $\partial_i b = pS(\partial_{i-1} g, \partial_{i-1} c) = pS(\tau(b_i), \partial_0 b_i) = b_i$ if $i > 1$.

LEMMA 21.4: If $G \times_\tau B$ is a PTCP of type (W) and G is a minimal complex, then $(E(\tau), p, B)$ is a minimal fibre space.

Proof: By Proposition 18.4, p is a minimal fibration. We must prove that B is a minimal complex. Thus suppose $b, c \in B_q$ and $\partial_i b = \partial_i c$, $i \neq k$. We must prove $\partial_k b = \partial_k c$. Let

$$x = (\tau(b), \partial_0 b) \text{ and } y = (\tau(c), \partial_0 c).$$

If $i + 1 \neq k$,

$$S(\partial_i x) = (e_{q-1}, \partial_{i+1} b) = (e_{q-1}, \partial_{i+1} c) = S(\partial_i y),$$

hence $\partial_i x = \partial_i y$. Therefore $\partial_i \tau(b) = \partial_i \tau(c)$ if $0 < i \neq k-1$ and $\tau(\partial_0 b)\partial_0 \tau(b) = \tau(\partial_0 c)\partial_0 \tau(c)$ if $k \neq 1$. Thus if $k > 0$, then $\partial_i \tau(b) = \partial_i \tau(c)$ for $i \neq k-1$, and, since G is minimal,

$$\partial_{k-1}\tau(b) = \partial_{k-1}\tau(c).$$

It follows that $\partial_k b = pS(\tau(\partial_k b), \partial_0\partial_k b) = \partial_k c$. If $k = 0$,

$$\partial_i \tau(b) = \partial_i \tau(c) \quad \text{for} \quad i > 0,$$

hence $\partial_0 \tau(b) = \partial_0 \tau(c)$ and therefore $\tau(\partial_0 b) = \tau(\partial_0 c)$. Then $\partial_0 b = pS(\tau(\partial_0 b), \partial_0\partial_0 b) = \partial_0 c$.

PROPOSITION 21.5: If $G \times_\tau B$ is a PTCP of type (W), then $E(\tau)$ is contractible.

Proof: Taking $\phi = (e_0, b_0)$ as base point, (i) and (ii) of Lemma 21.2 imply that $\partial\tilde{C}(S) + \tilde{C}(S)\partial$ is the identity map of

$\tilde{C}(E(\tau))$, and therefore $\tilde{H}_n(E(\tau)) = 0$ for all $n \geq 0$. Since $E(\tau)$ is a Kan complex, Corollary 13.7 implies that $E(\tau)$ is contractible provided that $\pi_1(E(\tau), \phi) = 0$. Suppose $x \in \tilde{E}(\tau)_1$, that is,

$$\partial_0 x = \phi = \partial_1 x.$$

Then $\partial_0 S(x) = x$, $\partial_1 S(x) = S(\partial_0 x) = S(\phi)$, and

$$\partial_2 S(x) = S(\partial_1 x) = S(\phi).$$

Clearly $S(\phi) = s_0 \phi$, and therefore $[x][s_0 \phi] = [s_0 \phi]$. Thus $[x] = 0$, as was to be proven.

Before proving the next, and fundamental, property of PTCP's of type (W), we need a definition.

DEFINITION 21.6: If G and G' are group complexes which operate on E and on E' from the right and if $\gamma: G \to G'$ is a simplicial homomorphism, then a simplicial map $\theta: E \to E'$ is said to be γ-equivariant if $\theta(xg) = \theta(x)\gamma(g)$ for all $x \in E_q$, $g \in G_q$. If $G = G'$ and $\gamma = 1_G$, θ is said to be equivariant.

THEOREM 21.7: Let $G \times_\tau B$ and $G' \times_{\tau'} B'$ be PTCP's and let $\gamma: G \to G'$ be a simplicial homomorphism. Then if $G' \times_{\tau'} B'$ is of type (W), there exists a unique γ-equivariant map $\theta: E(\tau) \to E(\tau')$ such that $\theta(e_q \times B_q) \subset e'_q \times B'_q$, $q \geq 0$.

Proof: θ is determined, if it exists, by its values on $e_q \times B_q$, $q \geq 0$, and since $B'_0 = b'_0$, $\theta(e_0, b) = (e'_0, b'_0)$ defines θ on $E(\tau)_0$. Again, if θ exists, then for $b \in b_q$, $q \geq 1$, there exists $\pi(b) \in B'_q$ such that $\theta(e_q, b) = (e_q, \pi(b)) = S\partial_0(e_q, \pi(b))$.

$$\partial_0 \theta(e_q, b) = \partial_0(e_q, \pi(b))$$

and, requiring $\partial_0 \theta = \theta \partial_0$, we find that θ is necessarily given by the inductive formula

(i) $\theta(e_q, b) = S\theta\partial_0(e_q, b)$, $b \in B_q$, $q \geq 1$.

It remains to prove that θ so defined is a simplicial map. $\partial_0 \theta = \theta \partial_0$ is true by construction and if $b \in B_1$, then

$$\partial_1 \theta(e_1, b) = (e_0', b_0') = \theta \partial_1 (e_1, b) .$$

Now suppose $\partial_i \theta = \theta \partial_i$, $0 \leq i \leq j$. Then we find

$$\partial_{j+1} \theta = \partial_{j+1} S \theta \partial_0 = S \partial_j \theta \partial_0 = S \theta \partial_0 \partial_{j+1} = \theta \partial_{j+1}$$

on $e_q \times B_q$, $q > 1$. Next, $S = s_0$ on $e_q \times B_q$, $q \geq 0$, hence $s_0 \theta = S \theta \partial_0 s_0 = \theta s_0$ on $e_q \times B_q$. Finally, assume $s_i \theta = \theta s_i$, $0 \leq i \leq j$. Then

$$s_{j+1} \theta = s_{j+1} S \theta \partial_0 = S \theta \partial_0 s_{j+1} = \theta s_{j+1}$$

on $e_q \times B_q$, $q \geq 1$.

COROLLARY 21.8: Any two PTCP's of type (W) with group G are naturally isomorphic.

Proof: If $E(\tau)$ and $E(\tau')$ are PTCP's with group G, there exist natural maps θ: $E(\tau) \to E(\tau')$ and θ': $E(\tau') \to E(\tau)$. By uniqueness, $\theta\theta'$ and $\theta'\theta$ are the respective identities.

For any simplicial group G, we now construct a PTCP $G \times_{\tau(G)} \overline{W}(G)$ which is of type (W). Let $\overline{W}_0(G)$ have one element $[\;]$ and define $\overline{W}_n(G) = G_{n-1} \times G_{n-2} \times \cdots \times G_0$, $n > 0$. Write elements of $\overline{W}_n(G)$ in the form $[g_{n-1}, \ldots, g_0]$, $g_i \in G_i$. Define face and degeneracy operators on $\overline{W}(G)$ by $s_0[\;] = [e_0]$, by

$$\partial_i [g_0] = [\;], \quad i = 0 \text{ or } 1,$$

and if $n \geq 1$ by

(i) $\partial_0 [g_n, \ldots, g_0] = [g_{n-1}, \ldots, g_0]$;

(ii) $\partial_{i+1} [g_n, \ldots, g_0] = [\partial_i g_n, \ldots, \partial_1 g_{n-i+1}, g_{n-i-1} \cdot \partial_0 g_{n-i}, g_{n-i-2}, \ldots, g_0]$;

(iii) $s_0 [g_{n-1}, \ldots, g_0] = [e_n, g_{n-1}, \ldots, g_0]$;

(iv) $s_{i+1}[g_{n-1},\ldots,g_0] =$

$$[s_i g_n,\ldots, s_0 g_{n-i}, e_{n-i}, g_{n-i-1},\ldots, g_0].$$

Define $\tau(G)$ on $\overline{W}_n(G)$, $n > 0$, by

(v) $\tau(G)[g_{n-1},\ldots,g_0] = g_{n-1}$.

We could verify directly that $\overline{W}(G)$ is a complex and that $\tau(G)$ is a twisting function, but the following inductive description of $\overline{W}(G)$ and $W(G)$ will make these facts obvious. Define an isomorphism of sets $\lambda\colon \overline{W}_{n+1}(G) \to W_n(G)$ by

(vi) $\lambda[g_n,\ldots,g_0] = (g_n, [g_{n-1},\ldots,g_0]),\quad n \geq 0$.

Let $\sigma(G)\colon \overline{W}_n(G) \to \overline{W}_n(G)$ be the canonical pseudo-cross section, $\sigma(G)(w) = (e_n, w)$. Now knowing the face and degeneracy operators on $\overline{W}_n(G)$, those on $W_n(G)$ are of course determined by the requirement that $W(G) = G \times_{\tau(G)} \overline{W}(G)$. Knowing the face and degeneracy operators on $W_n(G)$, those on $\overline{W}_{n+1}(G)$ are determined by:

(vii) $\partial_0 = p\lambda;\ \partial_{i+1} = \lambda^{-1}\partial_i\lambda;\quad s_0 = \lambda^{-1}\sigma(G);$

$s_{i+1} = \lambda^{-1}s_i\lambda$.

Finally, define $S = \sigma\lambda^{-1}\colon W_n(G) \to e_{n+1}\times\overline{W}_{n+1}(G)$, that is

(viii) $S(g_n, [g_{n-1},\ldots,g_0]) = (e_{n+1}, [g_n,\ldots,g_0])$.

Clearly $\partial_0 S$ is the identity of $W_n(G)$ and $S\partial_0$ is the identity on $e_{n+1}\times\overline{W}_{n+1}(G)$.

Given a simplicial homomorphism $\gamma\colon G \to G'$, define

(ix) $\overline{W}(\gamma)[g_n,\ldots,g_0] = [\gamma(g_n),\ldots,\gamma(g_0)]$.

With this definition, $\overline{W}$ becomes a functor from the category of simplicial groups to that of simplicial sets. $\overline{W}(G)$ will be called the classifying complex of G, $W(G)$ the universal G-bundle. We now proceed to justify this terminology.

LEMMA 21.9: Every PTCP $G \times_\tau B$ is induced from a simplicial map $\pi: B \to \overline{W}(G)$.

Proof: Let $\theta: E(\tau) \to W(G)$ be the equivariant map obtained in Theorem 21.7 and let $\pi: B \to \overline{W}(G)$ be defined by

$$\theta(e_q, b) = (e_q, \pi(b)) .$$

Then $\partial_0 \theta = \theta \partial_0$ implies $\tau = \tau(G) \circ \pi = \tau(G)^\pi$, so that $G \times_\tau B$ is induced from π and $\theta = \tilde{\pi}$. We remark that π is given explicitly by the formula:

(1) $\pi(b) = [\tau(b), \tau(\partial_0 b), \ldots, \tau(\partial_0^i b), \ldots, \tau(\partial_0^{n-1} b)]$, where

$\partial_0^i = \partial_0 \cdots \partial_0$, i factors of ∂_0.

The following remarks will lead to a simple proof of the classification theorem.

REMARKS 21.10: By Propositions 18.4 and 18.7, a PTCP is just a principal fibration with a given pseudo-cross section. From this point of view, a map $\pi: B \to \overline{W}(G)$ induces a principal fibration with given pseudo-cross σ^π, namely the canonical pseudo-cross section of $G \times_{\tau(G)^\pi} B$. σ^π is compatible with $\sigma(G)$ in the sense that $\tilde{\pi}\sigma^\pi = \sigma(G) \cdot \pi$. We define a morphism of principal fibrations with group G to be an equivariant simplicial map. Then a strong isomorphism of PTCP's with group G may be regarded as an automorphism of a single principal fibration.

LEMMA 21.11: Let $p: E \to B$ be a principal fibration with group G, let $B' \subset B$, and define $E' = p^{-1}(B')$. Then any morphism $\theta': E' \to W(G)$ may be extended to a morphism $\theta: E \to W(G)$.

Proof: θ' defines a pseudo-cross section $\sigma': B' \to E'$. By Lemma 18.6, σ' may be extended to $\sigma: B \to E$. By Proposition 18.7, σ defines a twisting function $\tau: B \to G$ and the resulting equivariant map $\theta: E = E(\tau) \to W(G)$ extends θ'.

THEOREM 21.12: $\pi \simeq \lambda: B \to \overline{W}(G)$ if and only if the PTCP's $W(G)^{\pi}$ and $W(G)^{\lambda}$ are strongly isomorphic.

Proof: If $\pi \simeq \lambda$, then $W(G)^{\pi}$ is strongly isomorphic to $W(G)^{\lambda}$ by Theorem 20.4. Conversely, suppose $W(G)^{\pi}$ and $W(G)^{\lambda}$ are strongly isomorphic. Regard $W(G)^{\pi}$ and $W(G)^{\lambda}$ as copies of the same principal fibration E. Let G operate on $E \times I$ by $(x, u)g = (xg, u)$, so that $p \times 1: E \times I \to B \times I$ is a principal fibration. Identify $W(G)^{\pi}$ with $E \times (0)$ and $W(G)^{\lambda}$ with $E \times (1)$. Then $E' = W(G)^{\pi} \cup W(G)^{\lambda}$ is a sub principal fibration of $E \times I$, and $\tilde{\pi}: W(G)^{\pi} \to W(G)$ and $\tilde{\lambda}: W(G)^{\lambda} \to W(G)$ together define a morphism $\theta': E' \to W(G)$. By the previous lemma, θ' can be extended to $\theta: E \times I \to W(G)$. If θ induces ψ on $B \times I$, then ψ is a homotopy from π to λ.

Combining Lemma 20.2, Proposition 20.7, and Lemma 21.9 with the theorem above, we obtain the classification theorem:

THEOREM 21.13: Let F be a complex on which G operates effectively from the left. Then the assignment to any map

$$\pi: B \to \overline{W}(G)$$

of the TCP (or G-bundle) with fibre F associated with the PTCP (or principal G-bundle) induced from π defines a one-to-one correspondence between the homotopy classes of maps $B \to \overline{W}(G)$ and the strong isomorphism classes of TCP's (or strong G-equivalence classes of G-bundles) with fibre F and base B.

REMARKS 21.14: Given a G-bundle $p: E \to B$, the corresponding homotopy class of maps $\pi: B \to \overline{W}(G)$ defines a homomorphism $\pi^*: H^*(\overline{W}(G), A) \to H^*(B, A)$. If A is a ring, π^* is a morphism of rings, and the image of π^* is called the characteristic subring of p.

For example, if $G = S(0(n))$ and $A = Z_2$ we obtain the Stiefel-Whitney classes, and if $G = S(U(n))$ and $A = Z$ we obtain the Chern classes.

Bibliographical Notes on Chapter IV

The material of section 17 is all due to Moore [52]. The definition of T C P adopted here is a special case of that of Moore [52], and gives what is called a regular T C P in [1]. The comparison of principal fibrations and P T C P's follows Cartan [6]. Of course, the simplicial concepts of principal fibration and principal bundle are equivalent, but the connection with topology is more easily seen using principal fibrations. Nearly all of the material of sections 19 and 20 is due to Barratt, Gugenheim, and Moore [1].

The complexes $\overline{W}(G)$ were first constructed (without the introduction of $W(G)$) by Eilenberg and MacLane [13]. The construction of $G \times_{\tau(G)} \overline{W}(G)$ is due to MacLane [40]. The axiomatic definition of P T C P's of type (W) is due to Moore [5, 52].

The classification theorem as stated here is developed in [1], but our proof follows Cartan [6].

CHAPTER V

EILENBERG-MACLANE COMPLEXES AND POSTNIKOV SYSTEMS

In this chapter, we first prove that the category of simplicial Abelian groups is isomorphic to that of chain complexes. We then introduce the minimal Abelian group complexes $K(\pi, n)$. A study of their properties leads to a proof that every Abelian group complex is of the homotopy type of a product of $K(\pi, n)$'s. We also obtain the well-known characterization of cohomology operations in terms of the cohomology of Eilenberg-MacLane complexes. Finally, we will obtain the k-invariants of Postnikov systems by means of a study of fibre bundles with fibre a $K(\pi, n)$.

§22. Simplicial Abelian groups

In this section, we investigate the category of simplicial Abelian groups. We will prove that this category is equivalent to that of (Abelian) chain complexes. An incidental result is the normalization theorem for simplicial Abelian groups. Finally, we will compare the homotopy and homology groups of a simplicial Abelian group.

We first develop an alternative definition of the homotopy groups of an Abelian group complex G. Let $A(G)$ denote G regarded as a chain complex with differential ∂ defined on G_n by $\partial = \sum_{i=0}^{n} (-1)^i \partial_i$. Let $N(G)$ denote the chain complex $\overline{G}$ defined in section 17, so that $N_n(G) = G_n \cap \operatorname{Ker} \partial_0 \cap \ldots \cap \operatorname{Ker} \partial_{n-1}$. For convenience, we redefine the differential ∂ on $N(G)$ by $\partial(x) = (-1)^n \partial_n$, $x \in N_n(G)$. Then $N(G)$ is a chain subcomplex of $A(G)$ and there results a natural homomorphism

$$i\colon \pi_n(G) = H_n(N(G)) \to H_n(A(G)).$$

THEOREM 22.1: Let G be an Abelian group complex. Then $i\colon \pi_n(G) \to H_n(A(G))$ is an isomorphism for all n.

Proof: Filter $A(G)$ by $x \in F^pA_n(G)$ if $x \in G_n$ and $\partial_i x = 0$ for $0 \leq i < \min(n, p)$. Then $F^pA(G) = A(G)$ if $p \leq 0$ and $F^pA_n(G) = N_n(G)$ if $p \geq n$, so that $\bigcap_p F^pA(G) = N(G)$. Each $F^{p+1}A(G)$ is a chain subcomplex of $F^pA(G)$. We will prove that each of the inclusions $i^p\colon F^{p+1}A(G) \subset F^pA(G)$ induces an isomorphism on homology. It will then follow that i is an isomorphism. We first define an epimorphism of chain complexes $f^p\colon F^pA(G) \to F^{p+1}A(G)$ by

$$f^p(x) = \begin{cases} x \quad \text{if} \quad x \in F^pA_n(G) & \text{where } n < p+1 \\ x - s_p\partial_p x \quad \text{if} \quad x \in F^pA_n(G) & \text{where } n \geq p+1. \end{cases}$$

Clearly $f^p(x) \in F^{p+1}A(G)$ and $f^p \circ i^p$ is the identity map of $F^{p+1}A(G)$. A simple calculation proves that $\partial f^p(x) = f^p(\partial x)$ for all x. We will complete the proof by obtaining a chain homotopy t^p between $i^p \circ f^p$ and the identity map of $F^pA(G)$. Thus define

$$t^p(x) = \begin{cases} 0 \quad \text{if} \quad x \in F^pA_n(G) & \text{where } n < p \\ (-1)^p s_p x \quad \text{if} \quad x \in F^pA_n(G) & \text{where } n \geq p. \end{cases}$$

Then $t^p(x) \in F^pA_{n+1}(G)$ and it is easily verified that $\partial t^p(x) + t^p\partial(x) = x - (i^p \circ f^p)(x)$, as desired.

COROLLARY 22.2: Let G be an Abelian group complex. Then $A(G) = N(G) \oplus D(G)$, where $D(G)$ is the chain subcomplex of $A(G)$ generated by the degenerate elements of G.

Proof: Let $i = i^0 \circ \ldots \circ i^{n-1} : N_n(G) \to A_n(G)$ and let $f = f^{n-1} \circ \ldots \circ f^0 : A_n(G) \to N_n(G)$. $f \circ i$ is the identity map on $N(G)$ and therefore $A(G) = N(G) \oplus \operatorname{Ker} f$. It suffices to prove $\operatorname{Ker} f = D(G)$, and $\operatorname{Ker} f \subset D_n(G)$ is clear from the definition of f. Let $x = \Sigma_{i=0}^{n-1} s_i y_i \in D(G)$. We must prove $f(x) = 0$, and we have

$$f(x) = f^{n-1} \circ \ldots \circ f^1(x - s_0\partial_0 x)$$
$$= f^{n-1} \circ \ldots \circ f^1(x^1) \qquad \text{where } x^1 = \sum_{i=1}^{n-1} s_i y_i^1,\ y_i^1 = y_i - s_0\partial_0 y_i$$
$$= \ldots = f^{n-1} \circ \ldots \circ f^j(x^j) \quad \text{where } x^j = \sum_{i=j}^{n-1} s_i y_i^j,\ y_i^j = y_i^{j-1} - s_{j-1}\partial_{j-1} y_i^{j-1}$$
$$= \ldots = f^{n-1}(s_{n-1} y_{n-1}^{n-1}) = 0, \text{ as claimed.}$$

COROLLARY 22.3: (Normalization Theorem). Let G be an Abelian group complex and let $\pi: A(G) \to A_N(G) = A(G)/D(G)$ be the projection. Then π is a chain equivalence.

Proof: Identifying $A_N(G)$ with $N(G)$, we find $\pi = f$. But $i \circ f: A(G) \to A(G)$ is a chain equivalence by the proof of the theorem.

N is a functor from the category $\mathcal{A}$ of simplicial Abelian groups to the category $\mathcal{C}$ of chain complexes. We now show that there exists a functor $\Gamma: \mathcal{C} \to \mathcal{A}$ such that $\Gamma \circ N$ and $N \circ \Gamma$ are the identity functors of $\mathcal{A}$ and $\mathcal{C}$. Thus let X be a chain complex. Define $\Gamma(X)$ as follows:

(i) $\Gamma_n(X) = X_n \oplus \sum_{r=0}^{n-1} \sum_{k=n-r} \sigma_{j_k} \ldots \sigma_{j_1} X_r$, where $\sigma_{j_k} \ldots \sigma_{j_1} X_r$ is the Abelian group whose elements are symbols $\sigma_{j_k} \ldots \sigma_{j_1} x$, $x \in X_r$, with the addition defined by

$$\sigma_{j_k} \ldots \sigma_{j_1} x + \sigma_{j_k} \ldots \sigma_{j_1} y = \sigma_{j_k} \ldots \sigma_{j_1}(x + y),$$

and where the sum is taken over all sequences $\{j_i\}$ such that $n > j_k > \ldots > j_1 \geq 0$.

(ii) $\partial_i: \Gamma_n(X) \to \Gamma_{n-1}(X)$ is defined by

(a) $\partial_n x = \partial(x)$ and $\partial_i x = 0$ if $i < n$, $x \in X_n$

(b) If $k = n - r$ and $x \in X_r$, then

$$\partial_i \sigma_{j_k} \cdots \sigma_{j_1} x = \begin{cases} \sigma_{h_{k-1}} \cdots \sigma_{h_1} x & s_{h_{k-1}} \cdots s_{h_1} \\ \sigma_{h_k} \cdots \sigma_{h_1} \partial(x) \text{ if } & s_{h_k} \cdots s_{h_1} \partial_r \\ 0 & s_{h_k} \cdots s_{h_1} \partial_j, j < r \end{cases} = \partial_i s_{j_k} \cdots s_{j_1},$$

when $\partial_i s_{j_k} \cdots s_{j_1}$ is expressed in the canonical form derived from formula (3) on page 4.

(iii) $s_i: \Gamma_n(X) \to \Gamma_{n+1}(X)$ is defined by

(a) $s_i x = \sigma_i x$, $x \in X_n$

(b) If $k = n - r$ and $x \in X_r$, then

$$s_i \sigma_{j_k} \cdots \sigma_{j_1} x = \sigma_{h_{k+1}} \cdots \sigma_{h_1} x \text{ if } s_i s_{j_k} \cdots s_{j_1} = s_{h_{k+1}} \cdots s_{h_1},$$

where $s_i s_{j_k} \cdots s_{j_1}$ is expressed in canonical form.

It is easily verified that $\Gamma(X)$ is a simplicial Abelian group. Now we can state

THEOREM 22.4: The functors $N: \mathfrak{A} \to \mathfrak{C}$ and $\Gamma: \mathfrak{C} \to \mathfrak{A}$ satisfy $\Gamma \circ N = 1_{\mathfrak{A}}$ and $N \circ \Gamma = 1_{\mathfrak{C}}$. Further, if $h: G \to G'$ is a simplicial homotopy, $h: f \simeq g$, then there exists a chain homotopy

$$s: N(G) \to N(G'), \quad s: N(f) \simeq N(g).$$

Conversely, if $s: X \to X'$ is a chain homotopy, $s: f \simeq g$, then there exists a simplicial homotopy $h: \Gamma(X) \to \Gamma(X')$, $h: \Gamma(f) \simeq \Gamma(g)$.

Proof: That $\Gamma \circ N = 1_{\mathfrak{A}}$ and $N \circ \Gamma = 1_{\mathfrak{C}}$ is an easy consequence of Corollary 22.2 and the construction of Γ. Given the homotopy $h: G \to G'$, $h: f \simeq g$, define $s = \sum_{i=0}^{n} (-1)^i h_i$ on $A_n(G)$. Then $s: A(f) \simeq A(g)$. $s(D(G)) \subset D(G')$ and, identifying $N(G)$ and $N(G')$ with $A_N(G)$ and $A_N(G')$, we obtain a chain homotopy $s: N(f) \simeq N(g)$ by passing to quotients. Conversely, let $s: X \to X'$ be a chain homotopy, $s: f \simeq g$. Then $\partial s + s\partial = f - g$. Define

$h: \Gamma(X) \to \Gamma(X')$ as follows:

(i) If $x \in X_n$, then

$h_n(x) = \sigma_n f(x) - \sigma_n s\partial(x) - s(x)$

$h_{n-1}(x) = \sigma_{n-1} f(x) - \sigma_n s\partial(x)$

$h_i(x) = \sigma_i f(x)$ if $i < n-1$.

(ii) h_i is defined inductively on the symbols $\sigma_{j_k} \ldots \sigma_{j_1} x$ by

$h_i(\sigma_{j_k} \ldots \sigma_{j_1} x) = \sigma_{j_k} h_{i-1}(\sigma_{j_{k-1}} \ldots \sigma_{j_1} x)$ if $j_k \leq i-1$

$h_i(\sigma_{j_k} \ldots \sigma_{j_1} x) = \sigma_{j_k+1} h_i(\sigma_{j_{k-1}} \ldots \sigma_{j_1} x)$ if $j_k > i-1$.

A simple calculation then proves that h is a homotopy from $\Gamma(f)$ to $\Gamma(g)$, as desired.

We have now proven that the homotopy theory of simplicial Abelian groups is equivalent to the homology theory of chain complexes.

Theorem 22.1 has the following corollary comparing the homology and homotopy groups of simplicial Abelian groups.

COROLLARY 22.5: Let G be an Abelian group complex. Then there exists a natural epimorphism of groups $j: \tilde{H}_n(G) \to \pi_n(G)$ such that $j \circ h$ is the identity map of $\pi_n(G)$, where h is the Hurewicz homomorphism. In particular, h is a monomorphism.

Proof: Using 0 as base point, $\tilde{C}_n(G)$ is the free Abelian group generated by the non-zero elements of G_n. Thus there is a natural epimorphism $\tilde{C}(G) \to A(G)$, and this induces

$$j^{\#}: \tilde{H}_n(G) \to H_n(A(G)).$$

Clearly $j^{\#} \circ h = i$, $i: \pi_n(G) \xrightarrow{\cong} H_n(A(G))$. Letting $j = i^{-1} \cdot j^{\#}$, the result follows.

We will later use the corollary above to prove that every Abelian group complex has the homotopy type of a product of Eilenberg-MacLane complexes.

COROLLARY 22.6: Let K be a complex. Regarding $C(K)$ as an Abelian group complex, we have $\pi_n(C(K)) = H_n(K)$ and $\pi_n(\tilde{C}(K)) = \tilde{H}_n(K)$.

This corollary states that in some sense homology is a special case of homotopy.

§23. Eilenberg-MacLane complexes

Recall that an Eilenberg-MacLane complex of type (π, n) is a Kan complex K such that $\pi_n(K, \phi) = \pi$ and $\pi_i(K, \phi) = 0$, $i \neq n$. Such a complex K will be called a $K(\pi, n)$ if it is minimal. We will prove the existence and essential uniqueness of $K(\pi, n)$'s.

LEMMA 23.1: Let π be a group. Define a group complex K by $K_n = \pi$, $n \geq 0$, and by letting each ∂_i and s_i on K_n be the identity map of π. Then K is a $K(\pi, 0)$ and any other $K(\pi, 0)$ L is naturally isomorphic to K.

Proof: Clearly K is a $K(\pi, 0)$. Now $K_0 = L_0 = \pi$ and the identity map $K_0 \to L_0$ induces a unique simplicial monomorphism $f\colon K \to L$. Suppose $f\colon K_n \to L_n$ is an epimorphism, $n \geq 0$, and let $y \in L_{n+1}$. If $f(x_i) = \partial_i y$, then $f(\partial_i x_j) = f(\partial_{j-1} x_i)$ implies $x_i = x_j = x$, say. Let $z = \partial_i y$, all i. Then $\partial_i s_0 z = \partial_i y$ for all i, hence since $\pi_{n+1}(L) = 0$, $s_0 z \sim y$. By minimality $s_0 z = y$, and thus $f(s_0 x) = y$.

Now observe that if G is an Abelian group complex, then if we regard $\overline{W}_n(G)$ as the group $G_{n-1} \times \cdots \times G_0$, $\overline{W}(G)$ becomes an Abelian group complex. Define $\overline{W}^n(G) = \overline{W}(\overline{W}^{n-1}(G))$ for all $n \geq 1$, $\overline{W}^0(G) = G$. The next theorem follows inductively from Lemma 21.4 and Proposition 21.5.

THEOREM 23.2: Let π be a group. Then $\overline{W}(K(\pi,0)$ is a $K(\pi, 1)$.

If π is Abelian, then the Abelian group complex $\overline{W}^n(K(\pi, 0))$ is a $K(\pi, n)$.

Before proving uniqueness, we need some preliminaries.

DEFINITIONS 23.3: Let (K, ϕ) be a Kan pair, where ϕ is the only vertex of K. Define a complex $P(K)$ by letting

$$\lambda \colon K_{n+1} \to P_n(K)$$

be an isomorphism of sets and by defining $\partial_i = \lambda \partial_{i+1} \lambda^{-1}$ and $s_i = \lambda s_{i+1} \lambda^{-1}$. Define $p\colon P_n(K) \to K_n$ by $p = \partial_0 \lambda^{-1}$ and let $L(K) = p^{-1}(\phi)$. It is easily verified that p is a Kan fibration and therefore $P(K)$ and $L(K)$ are Kan complexes. Let

$$S = \lambda s_0 \lambda^{-1} \colon P_n(K) \to P_{n+1}(K).$$

Then $\partial_0 S$ is the identity map of $P(K)$, $\partial_1 S(x) = \lambda(s_0 \phi)$ if $x \in P_0(K)$, and $\partial_{i+1} S = S \partial_i$ on $P_n(K)$, $n \geq 1$. As in the proof of Proposition 21.5, it follows that $P(K)$ is contractible.

LEMMA 23.4: Let (K, ϕ) be a Kan pair, where ϕ is the only vertex of K. Then $p\colon P(K) \to K$ is a PTCP of type (W) if $L(K)$ is a group complex.

Proof: Define $\sigma = \lambda s_0 \colon K_n \to P_n(K)$. Then $p\sigma$ is the identity map of K and we may identify $P_n(K)$ with $L_n(K) \times \sigma(K_n)$. An easy calculation proves that we may define an operation from the right of $L(K)$ on $P(K)$ by $(\ell_1, \sigma(k))\ell_2 = (\ell_1 \ell_2, \sigma(k))$. It follows that $p\colon P(K) \to K$ may be regarded as a principal fibration with group $L(K)$. Proposition 18.7 now implies that $P(K)$ is a PTCP. By definition $\partial_0 S$ is the identity of $P(K)$ and $S\partial_0$ is clearly the identity on $e_n \times \sigma(K_n)$. Therefore $p\colon P(K) \to L$ is of type (W).

REMARK 23.5: If K is a group complex, then so is $L(K)$, and therefore K is isomorphic to $\overline{W}(L(K))$. This implies the existence of a group complex structure on $\overline{W}(L(K))$. However, unless K is

Abelian, this product will not be that obtained by regarding $\overline{W}_n(L(K))$ as $L_{n-1}(K) \times \ldots \times L_0(K)$.

THEOREM 23.6: Let K and L be $K(\pi, n)$'s. Then there exists an isomorphism $K \to L$. Further, if π is Abelian, then any $K(\pi,n)$ has a structure of Abelian group complex.

Proof: It suffices to obtain an isomorphism $K \to L$, where $L = \overline{W}^n(K(\pi, 0))$. We proceed by induction on n and assume the result for $n' < n$, $n > 0$. Since K is a minimal complex, it is easy to see that $p: P(K) \to K$ is a minimal fibration. Therefore $L(K)$ is a $K(\pi, n-1)$. By the induction hypothesis, $L(K)$ is isomorphic to $\overline{W}^{n-1}(K(\pi, 0))$. But then $L(K)$ is a group complex and, by the lemma above and by Theorem 21.7, it follows that K is isomorphic to $\overline{W}^n(K(\pi, 0))$.

REMARKS 23.7: Observe that if π is an Abelian group and we define a chain complex $C(\pi, n)$ by $C_n(\pi, n) = \pi$ and $C_q(\pi, n) = 0$ for $q \neq n$, then $\Gamma C(\pi, n)$ is a $K(\pi, n)$. This gives an alternative proof of the existence of $K(\pi, n)$'s. Now, knowing that any $K(\pi, n)$ is an Abelian group complex, it follows that any $K(\pi, n)$ is equal to $\Gamma C(\pi, n)$. This is true since by Proposition 17.5 (recalling that what is there denoted $\overline{G}$ is the chain complex $N(G)$), we must have $NK(\pi, n) = C(\pi, n)$, and since $\Gamma NK(\pi, n) = K(\pi, n)$ by Theorem 22.4. As a corollary of these remarks, we have:

PROPOSITION 23.8: Let $f: \pi \to \pi'$ be a homomorphism of Abelian groups. Then there exists a unique simplicial homomorphism $\phi: K(\pi, n) \to K(\pi', n)$ such that $\phi = f: K(\pi, n)_n \to K(\pi', n)_n$.

When π is an Abelian group, there is an alternative, and more usual, description of $K(\pi, n)$. Let $C^*(\Delta[q], \pi)$ denote the normalized cochain complex and $C_*(\Delta[q])$ the normalized chain complex (previously denoted $C_N(\Delta[q])$, so that

$$C^*(\Delta[q], \pi) = \operatorname{Hom}_Z (C_*(\Delta[q]), \pi).$$

Fix an integer n and let $L(\pi, n+1)_q = C^n(\Delta[q], \pi)$, $q \geq 0$. Define a structure of Abelian group complex on $L(\pi, n+1)$ by

$$\partial_i u(x) = u(\delta_i x), \quad u \in L(\pi, n+1)_{q+1}, \quad x \in C_n(\Delta[q]),$$

and

$$s_i v(y) = v(\sigma_i y), \quad v \in L(\pi, n+1)_q, \quad y \in C_n(\Delta[q+1]),$$

where $\delta_i\colon C(\Delta[q]) \to C(\Delta[q+1])$ and $\sigma_i\colon C(\Delta[q+1]) \to C(\Delta[q])$ are induced from the simplicial maps $\delta_i\colon \Delta[q] \to \Delta[q+1]$ and $\sigma_i\colon \Delta[q+1] \to \Delta[q]$ defined in Definitions 5.4. Let

$$\overline{K}(\pi, n)_q = Z^n(\Delta[q], \pi),$$

the subgroup of $L(\pi, n+1)_q$ consisting of all cocycles. Observe that $\overline{K}(\pi, n)$ is a subcomplex of $L(\pi, n+1)$.

THEOREM 23.9: The Abelian group complex $\overline{K}(\pi, n)$ defined above is in fact a $K(\pi, n)$.

Proof: First observe that $\overline{K}(\pi, n)_q = 0$ if $q < n$ because every element of $\Delta[q]$ in dimension greater than q is degenerate. Next, $\overline{K}(\pi, n)_n = \pi$ since $Z^n(\Delta[n], \pi) = \operatorname{Hom}_Z (C_n(\Delta[n]), \pi)$ and $C_n(\Delta[n])$ is the free Abelian group generated by Δ_n. By Remarks 23.7, it suffices to prove that the chain complex $N\overline{K}(\pi, n)$ is $C(\pi, n)$. Clearly $N_n\overline{K}(\pi, n) = \overline{K}(\pi, n)_n = \pi$, and it remains to prove $N_q\overline{K}(\pi, n) = 0$, $q > n$. Thus let $\mu \in N_q\overline{K}(\pi, n)$, $q > n$, so that $\partial_i \mu = 0$, $0 \leq i \leq q-1$. Suppose for a contradiction that $\mu(x) \neq 0$, where $x = (a_0, \ldots, a_n)$, $0 \leq a_0 < \ldots < a_n \leq q$ is some basis element for $C_n(\Delta[q])$. Then $x \neq \delta_0 y$, since $\mu(\delta_0 y) = 0$, so $a_0 = 0$ (otherwise $x = \delta_0(a_0 - 1, \ldots, a_n - 1)$). Similarly $x \neq \delta_1 y$ implies $a_1 = 1$ and, inductively, $a_i = i$, $0 \leq i \leq n$. But μ is a cocycle, and this implies:

$$\delta(\mu)(0,1,\ldots,n+1) = \sum_{i=0}^{n+1} (-1)^i \mu(\partial_i(0,1,\ldots,n+1))$$

$$= \sum_{i=0}^{n} (-1)^i \mu(\delta_i \bar{x}) + (-1)^{n+1}\mu(x) = (-1)^{n+1}\mu(x) = 0,$$

where $\bar{x}$ denotes $x = (0,1,\ldots,n)$ regarded as an element of $\Delta[q-1]_n$. Thus $\mu(x) = 0$ and $\mu = 0$, as was to be proven.

From now on, the symbol $K(\pi, n)$ will denote the explicit complex $\overline{K}(\pi, n)$ described above. Now since $H^n(\Delta[q], \pi) = 0$, $\delta(C^n(\Delta[q], \pi)) = Z^{n+1}(\Delta[q], \pi)$. Thus the coboundaries δ define a simplicial homomorphism $\bar{\delta}\colon L(\pi, n+1) \to K(\pi, n+1)$. Further, addition defines an operation of $K(\pi, n)$ on $L(\pi, n+1)$ and of course $K(\pi, n) = \operatorname{Ker} \bar{\delta}$. It follows that $\bar{\delta}$ is a principal fibration with group $K(\pi, n)$. Define $\sigma\colon K(\pi, n+1)_q \to L(\pi, n+1)_q$ by

(i) $\sigma(\mu)(a_0, \ldots, a_n) = \mu(0, a_0, \ldots, a_n)$.

Then $\delta\sigma(\mu)(a_0, \ldots, a_{n+1}) = \sum_{i=0}^{n+1} (-1)^i \mu(0, a_0, \ldots, a_{i-1}, a_{i+1}, \ldots, a_{n+1})$

$$= -\mu(\partial(0, a_0, \ldots, a_{n+1})) + \mu(a_0, \ldots, a_{n+1})$$

$$= \mu(a_0, \ldots, a_{n+1}).$$

Thus $\delta\sigma$ is the identity of $K(\pi, n+1)$, and it is easily verified that σ is a pseudo-cross section of $\bar{\delta}$, the corresponding twisting function τ being explicitly defined by:

(ii) $\tau(\mu)(a_0, \ldots, a_n) = \mu(0, a_0+1, \ldots, a_n+1) - \mu(1, a_0+1, \ldots a_n+1)$.

Observing that $\nu \in \operatorname{Image}(\sigma)$ if and only if $\nu(a_0, \ldots, a_n) = 0$ whenever $a_0 = 0$, we define $S\colon L(\pi, n+1)_q \to \sigma(K(\pi, n+1))_{q+1}$ by

(iii) $S(\mu)(a_0, \ldots, a_n) = \mu(a_0 - 1, \ldots, a_n - 1)$ if $a_0 > 0$ and 0 if $a_0 = 0$.

Then $\partial_0 S$ is the identity on $L(\pi, n+1)$ and $S\partial_0$ is the identity on $\sigma(K(\pi, n+1))$. Thus $\bar{\delta}$ is a PTCP of type (W). Identifying $K(\pi, n)$ with $\overline{W}^n(K(\pi, 0))$ and letting $W^{n+1}(K(\pi, 0)) = W(\overline{W}^n(K(\pi, 0))$,

Theorem 21.7 implies:

THEOREM 23.10: The PTCP $\overline{\delta}: L(\pi, n+1) \to K(\pi, n+1)$ with group $K(\pi, n)$ is naturally isomorphic to the universal PTCP $p: W^{n+1}(K(\pi, 0)) \to \overline{W}^{n+1}(K(\pi, 0))$ of $K(\pi, n)$.

§24. $K(\pi, n)$'s and cohomology operations

We develop here certain fundamental properties of $K(\pi, n)$'s. These properties will lead to a proof that every Abelian group complex has the homotopy type of a product of $K(\pi, n)$'s and to the characterization of cohomology operations. In the next section, these properties will be used to define the k-invariants of Postnikov systems.

All chains and cochains mentioned in this section and the next are normalized.

Recall that $L(\pi, n+1)_n$ is π regarded as the group of homomorphisms of the free Abelian group generated by Δ_n into π. Define an n-cochain $u \in C^n(L(\pi, n+1), \pi)$ by $u(f) = f(\Delta_n)$, $f \in \mathrm{Hom}_Z(F(\Delta_n), \pi)$. u is called the fundamental cochain of $L(\pi, n+1)$.

LEMMA 24.1: Let $x \in L(\pi, n+1)_q$ and consider the simplicial map $\overline{x}: \Delta[q] \to L(\pi, n+1)$. If $\overline{x}^*: C^n(L(\pi, n+1), \pi) \to C^n(\Delta[q], \pi)$ is induced by $\overline{x}$, then $\overline{x}^*(u) = x$.

Proof: Since $C^n(\Delta[q], \pi) = L(\pi, n+1)_q$, the formula $\overline{x}^*(u) = x$ makes sense. From the definition of $L(\pi, n+1)$ as a complex and the requirement that $\overline{x}$ be a simplicial map, we derive $\overline{x}(y) = \overline{y}^*(x) \in C^n(\Delta[p], \pi)$ for $y \in \Delta[q]_p$, where $\overline{y}: \Delta[p] \to \Delta[q]$. In particular, if $y \in \Delta[q]_n$ is non-degenerate, then we must have:

$$\overline{x}^*(u)(y) = u(\overline{x}(y)) = \overline{x}(y)(\Delta_n) = \overline{y}^*(x)(\Delta_n) = x(\overline{y}(\Delta_n)) = x(y),$$

as was to be proven.

Let K be a complex and let $\text{Hom}(K, L(\pi, n+1))$ denote the set of simplicial maps $K \to L(\pi, n+1)$. The addition in $L(\pi, n+1)$ induces a structure of Abelian group on this set.

LEMMA 24.2: Let K be a complex. Define

$$\phi\colon \text{Hom}(K, L(\pi, n+1)) \to C^n(K, \pi)$$

by $\phi(f) = f^*(u)$. Then ϕ is an isomorphism of groups with inverse ψ defined by $\psi(\gamma)(x) = \bar{x}^*(\gamma)$ for $\gamma \in C^n(K, \pi)$, $x \in K_q$, where $\bar{x}^*\colon C^n(K, \pi) \to C^n(\Lambda[q], \pi)$.

Proof: We must first prove that $\psi(\gamma)$ is in fact a simplicial map. However, since $\partial_i = \delta_i^*$ and $s_i = \sigma_i^*$ on $L(\pi, n+1)$, we find:

$$\partial_i \psi(\gamma)(x) = \partial_i \bar{x}^*(\gamma) = \delta_i^* \bar{x}^*(\gamma) = (\bar{x}\delta_i)^*(\gamma) = (\overline{\partial_i x})^*(\gamma) = \psi(\gamma)(\partial_i x),$$

and similarly for the degeneracy operators.

ϕ and ψ are obviously homomorphisms, and it remains to prove that they are inverse to each other. If $x \in K_n$ is non-degenerate:

$$(\phi \circ \psi)(\gamma)(x) = \psi(\gamma)^*(u)(x) = u(\psi(\gamma)(x)) = u(\bar{x}^*(\gamma)) = \bar{x}^*(\gamma)(\Lambda_n) = \gamma(x).$$

Finally, if $x \in K_q$, then using the first lemma we find:

$$(\psi \circ \phi)(f)(x) = \bar{x}^*(\phi(f)) = \bar{x}^*(f^*(u)) = \overline{f(x)}^*(u) = f(x).$$

LEMMA 24.3: ϕ and ψ define an isomorphism between the cocycles $Z^n(K, \pi)$ and the maps $\text{Hom}(K, K(\pi, n))$.

Proof: If $\gamma \in Z^n(K, \pi)$, we must prove that $\psi(\gamma)(x) \in K(\pi, n)$ for all $x \in K_q$, that is, $\psi(\gamma)(x)$ is a cocycle. But

$$\delta(\psi(\gamma)(x)) = \delta(\bar{x}^*(\gamma)) = \bar{x}^*(\delta(\gamma)) = 0.$$

Conversely, suppose that $\psi(\gamma)(x)$ is a cocycle for all $x \in K_q$. Then if $x \in K_{n+1}$ is non-degenerate, we find that:

$$\delta\gamma(x) = \delta\gamma(\bar{x}(\Lambda_{n+1})) = \delta(\bar{x}^*(\gamma))(\Lambda_{n+1}) = \delta(\psi(\gamma)(x))(\Lambda_{n+1}) = 0,$$

and γ is a cocycle.

ϕ: $\mathrm{Hom}(K,K(\pi,n)) \to Z^n(K,\pi)$ is defined by $\phi(f) = f^*(u)$, where u is regarded as an element of $Z^n(K(\pi,n),\pi)$. Here u is called the fundamental cocycle of $K(\pi,n)$ and its cohomology class is called the fundamental class. We note that the following diagram commutes:

$$\begin{array}{ccc} \mathrm{Hom}(K,L(\pi,n+1)) & \xrightarrow{\phi} & C^n(K,\pi) \\ \big\downarrow \mathrm{Hom}(1,\overline{\delta}) & & \big\downarrow \delta \\ \mathrm{Hom}(K,K(\pi,n+1)) & \xrightarrow{\phi} & Z^{n+1}(K,\pi). \end{array}$$

Further, we have the following theorem.

THEOREM 24.4: Let $f,g \in \mathrm{Hom}(K,K(\pi,n))$. Then $f \simeq g$ if and only if $\phi(f)$ is cohomologous to $\phi(g)$. Thus ϕ induces an isomorphism of $\pi(K,K(\pi,n))$, the group of homotopy classes of maps $K \to K(\pi,n)$, onto $H^n(K,\pi)$.

Proof: If $f \simeq g$, then $\phi(f) = f^*(u) = g^*(u) = \phi(g)$ on the cohomology level, hence the cocycles $f^*(u)$ and $g^*(u)$ are cohomologous. Conversely, suppose $\phi(f) = \phi(g) + \delta(\alpha)$, $\alpha \in C^{n-1}(K,\pi)$. Let $i_0: K \to K \times (0)$ and $i_1: K \to K \times (1)$ be the simplicial maps identifying K with the cited subcomplexes of $K \times I$. It suffices to find $\gamma \in Z^n(K \times I, \pi)$ such that

$$i_0^*(\gamma) = \gamma | C^n(K \times (0)) = \phi(f)$$

and $i_1^*(\gamma) = \gamma | K \times (1) = \phi(g)$, for then $\psi(\gamma): K \times I \to K(\pi,n)$ will be a homotopy from f to g. Let $p: K \times I \to K$ be the projection onto K and let $\gamma_0 = p^*(\phi(f)) \in Z^n(K \times I,\pi)$. Clearly $i_1^*(\gamma_0) = i_0^*(\gamma_0) = \phi(f)$. Further, regarding α as a cochain defined on $i_1(K)$, we may choose a cochain $\beta \in C^{n-1}(K \times I, \pi)$ which extends α and vanishes on $i_0(K)$. Thus $i_0^*(\beta) = 0$ and $i_1^*(\beta) = \alpha$. Let $\gamma = \gamma_0 - \delta(\beta)$. Then $i_0^*(\gamma) = \phi(f)$ and $i_1^*(\gamma) = \phi(f) - \delta(\alpha) = \phi(g)$, as desired.

Before studying cohomology operations, we prove:

THEOREM 24.5: Let G be a connected Abelian group complex, and let $\pi_n = \pi_n(G)$. Then G has the homotopy type of the infinite Cartesian product $\times_{n=1}^{\infty} K(\pi_n, n)$.

Proof: By Corollary 22.5, there exists an epimorphism $j: H_n(G) \to \pi_n(G)$ such that $j \circ h: \pi_n \to \pi_n$ is the identity map, $n \geq 1$. Let $\gamma^n \in Z^n(G, \pi_n)$ satisfy $\gamma^n(x) = j\{x\}$, $x \in C_n(G)$, where $\{x\}$ denotes the homology class of the non-bounding cycle x. By Lemma 24.3, γ^n determines a map $f^n: G \to K(\pi_n, n)$. If $g \in G_n$ regarded as $A(G)_n$, then $f^n(g) = \overline{g}^*(\gamma^n)$ and

$$f^n(g)(\Delta_n) = \gamma^n(g) = j\{g\} = j \circ h[g] = [g].$$

Therefore $f^n_*: \pi_n(G) \cong \pi_n(K(\pi_n, n))$. Let $f = \times_{n=1}^{\infty} f^n: G \to \times_{n=1}^{\infty} K(\pi_n, n)$. Clearly f is a weak homotopy equivalence, hence by Theorem 12.5 a homotopy equivalence.

Now we define and characterize cohomology operations.

DEFINITION 24.6: Let π and π' be Abelian groups, n and n' integers. Regard $H^n(K, \pi)$ and $H^{n'}(K, \pi')$ as defining functors from the category of simplicial sets to that of sets. Then a cohomology operation of type (n, n', π, π') is a natural transformation of functors $C: H^n(\ ;\pi) \to H^{n'}(\ ;\pi')$.

Let $H^{n'}(\pi,n,\pi')$ denote the cohomology group $H^{n'}(K(\pi,n);\pi')$. Then we have:

THEOREM 24.7: The cohomology operations of type (n, n', π, π') are in natural one-to-one correspondence with the elements of $H^{n'}(\pi,n,\pi')$.

Proof: If C is a cohomology operation of type (n, n', π, π'), define $\alpha(C) = C(u) \in H^{n'}(\pi,n,\pi')$, where u denotes the fundamental class of $H^n(\pi,n,\pi)$. Conversely, suppose $c \in H^{n'}(\pi, n, \pi')$ and $\gamma \in H^n(K, \pi)$, where K is some complex. By Theorem 24.4, γ defines a homotopy class of maps $\psi(\gamma)\colon K \to K(\pi, n)$. Define a cohomology operation $\beta(c)$ by $\beta(c)(\gamma) = \psi(\gamma)^*(c)$. Then since $\psi(u)$ is the class of the identity map $K(\pi, n) \to K(\pi, n)$, we find

$$(\alpha \circ \beta)(c) = \beta(c)(u) = \psi(u)^*(c) = c.$$

Finally, if $\gamma \in H^n(K, \pi)$ for some complex K, then using the naturality of cohomology operations and the fact that $\psi(\gamma)^*(u) = \gamma$ we find

$$(\beta \circ \alpha)(C)(\gamma) = \beta(C(u))(\gamma) = \psi(\gamma)^*(C(u)) = C(\psi(\gamma)^*(u)) = C(\gamma).$$

§25. The k-invariants of Postnikov systems

Let (K, ϕ) be a connected Kan pair. We wish to obtain invariants which characterize the homotopy type of K. Choosing a minimal subcomplex if necessary, we may assume that K is minimal (by Theorem 9.5). Let $X = (X^{(0)}, \ldots, X^{(n)}, \ldots)$ denote the natural Postnikov system of K (Definition 8.5). Then by Lemma 12.1, each $X^{(n)}$ is a minimal fibre space. Letting π_n denote $\pi_n(K, \phi)$, the fibre $F^{(n+1)}$ of $X^{(n)}$ is isomorphic to $K(\pi_{n+1}, n+1)$ (by Lemma 10.2 and Theorem 23.6). Observe that $K^{(0)} = \phi$, so that $K^{(1)} \cong K(\pi_1, 1)$. Now by Theorem 11.11, each $X^{(n)}$ is a fibre bundle, $n \geq 1$. We will show that, under assumptions on π_1, the group of the bundle $X^{(n)}$ can be reduced to $K(\pi_{n+1}, n+1)$, $n \geq 1$. Then by Theorem 19.4, it will follow that $X^{(n)}$ can be given a structure of PTCP. By Lemma 21.9, there will result a map $f^{n+2}: K^{(n)} \to \overline{W}(K(\pi_{n+1}, n+1)) = K(\pi_{n+1}, n+2)$ which will induce the PTCP $X^{(n)}$. By Lemma 24.3, f^{n+2} will determine a cocycle $k^{n+2} \in Z^{n+2}(K^{(n)}, \pi_{n+1})$. A sequence k^{n+2}, $1 \leq n$, of cocycles so chosen is called a set of k-invariants of K, and such a sequence certainly determines the homotopy type of K.

Of course a similar argument holds for fibre spaces. Thus let $(F, \psi) \xrightarrow{i} (E, \psi) \xrightarrow{p} (B, \phi)$ be a fibre sequence of connected Kan pairs. By Corollary 10.11, we may assume that (E, p, B) is a minimal fibre space. Let $\mathcal{E} = \{\mathcal{E}^{(0)}, \ldots, \mathcal{E}^{(n)}, \ldots\}$ denote the natural Postnikov system of (E, p, B) (Definition 8.10). Lemmas 10.3 and 12.6 imply that each $\mathcal{E}^{(n)}$ is a minimal fibre space. Letting π_n denote $\pi_n(F, \psi)$, the fibre of $\mathcal{E}^{(n)}$ is isomorphic to $K(\pi_{n+1}, n+1)$, $n \geq 0$. Under assumptions on π_1, $\pi_1(E)$, and $\pi_1(B)$, we will show that the group of each of the fibre bundles $\mathcal{E}^{(n)}$ can be reduced to $K(\pi_{n+1}, n+1)$. Then each $\mathcal{E}^{(n)}$ will have a structure of PTCP, there will be a map $f^{n+2}: E^{(n)} \to \overline{W}(K(\pi_{n+1}, n+1)) = K(\pi_{n+1}, n+2)$ which will induce the PTCP $\mathcal{E}^{(n)}$, and there will result a cocycle $k^{n+2} \in Z^{n+2}(E^{(n)}, \pi_{n+1})$. A sequence k^{n+2}, $0 \leq n$, of cocycles so chosen is called a set of k-invariants of p, and such a sequence determines the strong homotopy type of (E, p, B).

To validate the arguments above, we must study fibre bundles with fibre $K(\pi, n)$. The group of any such bundle is contained in $A(K(\pi, n))$, the group complex consisting of invertible elements of the monoid complex $M(\pi, n) = K(\pi, n)^{K(\pi, n)}$. We determine the structure of $M(\pi, n)$ and $A(K(\pi, n))$ in the next two propositions.

We will need the following lemma.

LEMMA 25.1: The group of simplicial maps $\mathrm{Hom}(K(\pi, n), K(\pi', n))$ is isomorphic to the group of homomorphisms $\mathrm{Hom}_Z(\pi, \pi')$, and every element of $\mathrm{Hom}(K(\pi, n), K(\pi', n))$ is a homomorphism.

Proof: We observe first that every element

$$f \in \mathrm{Hom}(K(\pi, n), K(\pi', n))$$

is a homomorphism. This is true since f induces a homomorphism $\pi \to \pi'$ on $K(\pi, n)_n = \pi_n(K(\pi, n)) = \pi$ and since if $x, y \in K(\pi, n)_q$, $q > n$, then, by induction on q, $f(x+y)$ and $f(x) + f(y)$ have the same faces, hence are homotopic and therefore equal. Now f determines a homomorphism $\pi \to \pi'$ and the converse was proven in Proposition 23.8.

PROPOSITION 25.2: As a complex, $M(\pi, n)$ is isomorphic to $\mathrm{End}(\pi, \pi) \times K(\pi, n)$, where $\mathrm{End}(\pi, \pi)$ is the complex defined by $\mathrm{End}(\pi, \pi)_q = \mathrm{Hom}(K(\pi, n), K(\pi, n))$ for all q, with each ∂_i and s_i the identity map.

Proof: $M(\pi, n)_q = \mathrm{Hom}(K(\pi, n) \times \Delta[q], K(\pi, n))$. Define $L_q = Z^n(K(\pi, n) \times \Delta[q], \pi)$. Then ϕ of Lemma 24.3 defines an isomorphism $M(\pi, n)_q \to L_q$. Anticipating the Eilenberg-Zilber theorem, to be proven later, we know that the chain complexes

$$C_*(K(\pi, n) \times \Delta[q])$$

and $C_*(K(\pi, n)) \otimes C_*(\Delta[q])$ are chain homotopic. Further, since $K(\pi, n)_q = 0$, $q < n$, and 0 is a degenerate simplex, $q > 0$, we find:

(i) $C_n(K(\pi,n) \times \Delta[q]) \cong C_n(K(\pi,n)) \otimes C_0(\Delta[q]) \oplus C_0(K(\pi,n)) \otimes C_n(\Delta[q])$,

(ii) $B_n(K(\pi,n) \times \Delta[q]) \cong B_n(K(\pi,n)) \otimes C_0(\Delta[q]) \oplus C_n(K(\pi,n)) \otimes B_0(\Delta[q])$
$\oplus C_0(K(\pi,n)) \otimes B_n(\Delta[q])$.

Now suppose $\gamma \in L_q$, so that $\delta(\gamma) = 0$. From the second term on the right of (ii), we find $\gamma(x, y) = \gamma(x, z)$, where $x \in C_n(K(\pi, n))$ and $y, z \in \Delta[q]_0$. Then we see that if $\gamma = \gamma_1 + \gamma_2$, where

$$\gamma_1 = \gamma \mid C_n(K(\pi, n)) \otimes C_0(\Delta[q])$$

and $\gamma_2 = \gamma \mid C_0(K(\pi, n)) \otimes C_n(\Delta[q])$, we may regard γ_1 as an element of $Z^n(K(\pi, n), \pi)$ and γ_2 as an element of $Z^n(\Delta[q], \pi) = K(\pi, n)_q$. Translating back to $M(\pi, n)$, we obtain the result.

The result above determines the structure of $M(\pi, n)$ as an Abelian group complex, but we must still determine the monoid structure and identify $A(K(\pi, n)) \subset M(\pi, n)$.

PROPOSITION 25.3: The product in $M(\pi, n)$ is given by

(i) $(f, x)(g, y) = (fg, f(y) + x)$, $f, g \in \mathrm{End}(\pi, \pi)_q$, $x, y \in K(\pi, n)_q$.

Let $\mathrm{Aut}(\pi,\pi)_q \subset \mathrm{End}(\pi,\pi)_q$ denote the group of automorphisms $K(\pi, n) \to K(\pi, n)$. Then $(f, x) \in A(K(\pi, n))$ if and only if $f \in \mathrm{Aut}(\pi,\pi)$, and in that case:

(ii) $(f, x)^{-1} = (f^{-1}, -f^{-1}(x))$, $f \in \mathrm{Aut}(\pi,\pi)_q$, $x \in K(\pi, n)_q$.

Finally, the embedding $v\colon K(\pi, n) \to A(K(\pi, n))$ of $K(\pi, n)$ as a subgroup of operators on $K(\pi, n)$ is given by:

(iii) $v(x) = (1, x)$, $x \in K(\pi, n)_q$, where 1 denotes the identity map of $K(\pi, n)$.

Proof: If (f, x) and (g, y) are as in (i) and $z \in K(\pi, n)_r$, $w \in \Delta[q]_r$, then using Lemma 25.1 we find:

$$(f, x)\cdot(g, y)[(z, w)] = (f, x)[(g(z) + \overline{y}(w), w)] = f \cdot g(z) + f(\overline{y}(w)) + \overline{x}(w),$$
$$\text{where } \overline{x}, \overline{y}\colon \Delta[q] \to K(\pi, n) \text{ and } f(\overline{y}(w)) = \overline{f(y)}(w).$$

This implies (i) and (i) implies (ii). If $w = \varepsilon\Delta_q$, then:

$$v(x)(z, w) = z + \varepsilon(x) = z + \overline{x}(w) = (1, x)[(z, w)].$$

This proves (iii).

In the next two theorems, we obtain conditions on a bundle

with fibre $K(\pi, n)$ which guarantee that the group of the bundle can be reduced to $K(\pi, n)$ and show by how much two $K(\pi, n)$-bundles so obtained can differ.

THEOREM 25.4: Let $p: E \to B$ be a fibre bundle with fibre $K(\pi,n)$ and assume B is a connected Kan complex such that there exist no non-trivial homomorphisms of $\pi_1(B, \phi)$ into the group of automorphisms of π. Then the group of the bundle can be reduced to $K(\pi, n)$.

Proof: Let $\{a(b)\}$ be an $A(K(\pi, n))$-atlas of p with twisting function $\tau(b) = (f(b), x(b))$. Let $c \in B_2$. Then from

$$\partial_0 \tau(c) = \tau(\partial_0 c)^{-1} \tau(\partial_1 c)$$

and $\partial_1 \tau(c) = \tau(\partial_2 c)$ we derive $f(\partial_2 c) = f(c) = f(\partial_0 c)^{-1} f(\partial_1 c)$. Considering f as a function $B_1 \to \mathrm{Aut}(\pi, \pi)$, we find that $b \sim b'$ implies $f(b) = f(b')$ and that f induces a homomorphism

$$\pi_1(B, \phi) \to \mathrm{Aut}(\pi, \pi)$$

for any base point ϕ. Therefore $f(b) = 1$ for $b \in B_1$. Define $\overline{\beta}(b) = \beta(b)(g(b), 0)$, where $g(b) = 1$ for $b \in B_0$ and

$$g(b) = f^{-1}(b) f^{-1}(\partial_0 b) \ldots f^{-1}(\partial_0^{q-1} b) \text{ for } b \in B_q, \quad q > 0.$$

The relations $\partial_0 \tau(b) = \tau(\partial_0 b)^{-1} \tau(\partial_1 b)$ and $\partial_i \tau(b) = \tau(\partial_{i+1} b)$, $i > 0$, together with the condition $f(b) = 1$ for $b \in B_1$ are easily seen to imply that $g(b) = g(\partial_i b)$ for $i > 0$. Then:

(i) $\partial_i \overline{\beta}(b) = \partial_i \beta(b)(g(b),0) = \beta(\partial_i b)(g(\partial_i b),0) = \overline{\beta}(\partial_i b), \quad i > 0,$

and

(ii) $$\begin{aligned} \partial_0 \overline{\beta}(b) &= \beta(\partial_0 b)(f(b), x(b))(g(b),0) = \beta(\partial_0 b)(g(\partial_0 b), x(b)) \\ &= \beta(\partial_0 b)(g(\partial_0 b),0)(1, g(\partial_0 b)^{-1}(x(b))) \\ &= \overline{\beta}(\partial_0 b)(1, g(\partial_0 b)^{-1}(x(b))). \end{aligned}$$

Thus $\{a(b)\}$ is $A(K(\pi, n))$-equivalent to the $K(\pi,n)$-atlas $\{\overline{a}(b)\}$, and the result is proven.

THEOREM 25.5: Let $p: E \to B$ be a fibre bundle with fibre $K(\pi,n)$, and suppose the group of bundle can be reduced to $K(\pi,n)$. Then if B_0 has just one element, the resulting $K(\pi,n)$-bundle is unique up to an automorphism of π (in a sense to be made precise in the proof).

Proof: Let $\{\alpha(b)\}$ and $\{\overline{\alpha}(b)\}$ be two $K(\pi,n)$-atlases for p with twisting functions $\tau(b) = (1, x(b))$ and $\overline{\tau}(b) = (1, y(b))$. $\{\alpha(b)\}$ and $\{\overline{\alpha}(b)\}$ are necessarily $A(K(\pi,n))$-equivalent, say

$$\overline{\beta}(b) = \beta(b)(g(b), z(b)).$$

Then:

(i)
$$\begin{aligned}\partial_0 \overline{\beta}(b) &= \beta(\partial_0 b)(1, x(b))(g(b), \partial_0 z(b)) \\ &= \overline{\beta}(\partial_0 b)[(g(\partial_0 b), z(\partial_0 b))]^{-1}(1, x(b))(g(b), \partial_0 z(b)) \\ &= \overline{\beta}(\partial_0 b)(1, y(b)).\end{aligned}$$

It follows that $g(\partial_0 b)^{-1} g(b) = 1$, $g(b) = g(\partial_0 b)$. Since B_0 has just one element, we find inductively that $g(b)$ is the same element g for every b. In this sense, any two $K(\pi,n)$-atlases of p differ at most by an automorphism of π. In fact, replacing $\{\overline{\alpha}(b)\}$ by the $K(\pi,n)$-equivalent atlas $\{\tilde{\alpha}(b)\}$ where $\tilde{\beta}(b) = \overline{\beta}(b)(1, -g^{-1}(z(b)))$, we have:

(ii) $\tilde{\beta}(b) = \beta(b)(g, z(b))(1, -g^{-1}(z(b))) = \beta(b)(g, 0)$.

The resulting twisting functions may be inequivalent, but

$$(g^{-1}, 0)\tau(b)(g, 0) = \overline{\tau}(b)$$

or, considering τ and $\tilde{\tau}$ to take values in $v^{-1}(1, K(\pi,n))$,

$$g^{-1}(\tau(b)) = \tilde{\tau}(b).$$

REMARKS 25.6: Let $\tau: B \to K(\pi,n)$ be a twisting function and let $f: B \to \overline{W}(K(\pi,n)) = K(\pi, n+1)$ be the corresponding map. Then $\phi(f) \in Z^{n+1}(B,\pi)$ is given explicitly by $\phi(f)(b) = \tau(b)(\Delta_n)$, $b \in B_{n+1}$. This follows from a calculation showing that $\tau(K(\pi,n))(\psi\phi)(f) = \tau$

when $\phi(f)$ is so defined (where $\tau(K(\pi, n))$ is the twisting function for $W(K(\pi, n))$ defined by formula (ii) on page 102).
Therefore, if $k, \bar{k} \in Z^{n+1}(B, \pi)$ arise from twisting functions $\tau, \bar{\tau}$ such that $\bar{\tau} = g \cdot \tau$, where g is an automorphism of $K(\pi, n)$, then $\bar{k} = g \cdot k$, where g is regarded as an automorphism of π. Now we extend the concept of a map of PTCP's with fixed fibre and base to include g-equivariant maps, or g-maps, g an automorphism of the group. Then two cocycles k and $\bar{k}$ in $Z^{n+1}(B, \pi)$ determine g-isomorphic PTCP's with fibre $K(\pi, n)$ if and only if $g_*\{k\} = \{\bar{k}\}$, where $g_*: H^{n+1}(B, \pi) \to H^{n+1}(B, \pi)$ is induced by the automorphism g of π.

Now we have proven the following theorems:

THEOREM 25.7: Let (K, ϕ) be a connected minimal Kan pair and let π_n denote $\pi_n(K, \phi)$. Suppose there exist no non-trivial homomorphisms $\pi_1 \to \operatorname{Aut}(\pi_n, \pi_n)$, $n > 1$. Then each of the fibre bundles $X^{(n)}$, $n \geq 1$, of the natural Postnikov system of K is a $K(\pi_{n+1}, n+1)$-bundle, unique up to g-isomorphism,

$$g \in \operatorname{Aut}(\pi_{n+1}, \pi_{n+1}).$$

$X^{(n)}$ is determined by $k^{n+2} = \phi(f^{n+2}) \in Z^{n+2}(K^{(n)}, \pi_{n+1})$, where $f^{n+2}: K^{(n)} \to K(\pi_{n+1}, n+2)$ is defined by any twisting function of $X^{(n)}$. Another cocycle $\bar{k}^{n+2}$ also induces $X^{(n)}$ if and only if $\{\bar{k}^{n+2}\} = g_*\{k^{n+2}\}$ for some $g \in \operatorname{Aut}(\pi_{n+1}, \pi_{n+1})$.

THEOREM 25.8: Let $(F, \psi) \to (E, \psi) \xrightarrow{p} (B, \phi)$ be a fibre sequence of connected Kan pairs, where (E, p, B) is a minimal fibre space. Let π_n denote $\pi_n(F, \psi)$ and assume π_1 is Abelian. Suppose there exist no non-trivial homomorphisms $\pi_1(E, \psi) \to \operatorname{Aut}(\pi_n, \pi_n)$, $n \geq 1$. Then each of the fibre bundles $\mathcal{E}^{(n)}$, $n \geq 0$, of the natural Postnikov system of p is a $K(\pi_{n+1}, n+1)$-bundle, unique up to

g-isomorphism, $g \in \mathrm{Aut}(\pi_{n+1}, \pi_{n+1})$. $\mathcal{E}^{(n)}$ is determined by

$$k^{n+2} = \phi(f^{n+2}) \in Z^{n+2}(E^{(n)}, \pi_{n+1}),$$

where $f^{n+2}\colon E^{(n)} \to K(\pi_{n+1}, n+2)$ is defined by any twisting function of $\mathcal{E}^{(n)}$. Another cocycle $\bar{k}^{n+2}$ also induces $\mathcal{E}^{(n)}$ if and only if $\{\bar{k}^{n+2}\} = g_*\{k^{n+2}\}$ for some $g \in \mathrm{Aut}(\pi_{n+1}, \pi_{n+1})$.

REMARKS 25.9: There are various comments to be made on the arguments above:

(1) The k-invariants could be defined and should be interpreted as obstructions to the cross-sectioning of the relevant bundles.

(2) Had we developed the theory of the operation of $\pi_1(K, \phi)$ on $\pi_n(K, \phi)$ (and of $\pi_1(B, \phi)$ on $\pi_n(F, \psi)$), then the assumption as to non-existence of homomorphisms could have been replaced by the assumption of n-simplicity for all n. Further, the appropriate generalization to the case where n-simplicity fails could be developed by use of the results on the structure of $A(K(\pi, n))$'s.

(3) If (K, ϕ) is a Kan pair and π is an Abelian group such that $\mathrm{Aut}(\pi, \pi)$ is an Abelian group, then the statement that there exist no non-trivial homomorphisms $\pi_1(K, \phi) \to \mathrm{Aut}(\pi, \pi)$ is precisely the statement that $H^1(K, \mathrm{Aut}(\pi, \pi)) = 0$. This follows from the universal coefficient theorem and the fact that

$$H_1(K; Z) \cong \pi_1(K, \phi)/[\pi_1(K, \phi), \pi_1(K, \phi)].$$

(4) There is an obvious inverse procedure for constructing any (simple) homotopy type: Let there be given a sequence of groups $(\pi_1, \ldots, \pi_n, \ldots)$, π_n Abelian if $n > 1$. Define

$$K^{(1)} = K(\pi_1, 1).$$

Choose $k^3 \in Z^3(K^{(1)}, \pi_2)$, and construct $K^{(2)}$ as the PTCP induced from $\psi(k^3)\colon K^{(1)} \to K(\pi_2, 3)$. Choose $k^4 \in Z^4(K^{(2)}, \pi_3)$ and construct $K^{(3)}$ from $\psi(k^4)$, etc. Similarly, suppose π_1 is Abelian and let B be a connected minimal complex. Define $E^{(0)} = B$. Choose $k^2 \in Z^2(E^{(0)}, \pi_1)$ and let $E^{(1)}$ be induced from

$$\psi(k^2): E^{(0)} \to K(\pi_1, 2).$$

Choose $k^3 \in Z^3(E^{(1)}, \pi_2)$, and construct $E^{(2)}$ from $\psi(k^3)$, etc.

(5) The precise statement of the topological results implied by the arguments above should be clear. Starting with a connected CW-complex X or a Serre fibration $p: E \to B$ of connected CW-complexes, we derive (via the functor S, minimalization, and the functor T) a sequence of principal fibrations (in the topological sense) which are determined by k-invariants and which determine the homotopy type of X or of (E, p, B). The additional assumption that all homotopy groups in sight are countable must here be made unless one takes the precaution of working in the category of compactly generated spaces.

BIBLIOGRAPHICAL NOTES ON CHAPTER V

The result that $\pi_n(N(G)) = H_n(A(G))$ and the resulting comparison of the homotopy and homology of Abelian group complexes is due to Moore [52], but our proof is essentially that of Cartan [6]. The equivalence of the categories of simplicial Abelian groups and of chain complexes is due to Dold [8] and Kan [27]. Our construction of the functor Γ follows MacLane [41]. The relation

$$\pi_n(\tilde{C}(K)) = \tilde{H}_n(K)$$

is the simplicial analog of a result of Dold and Thom [9] on the infinite "symmetric power" of a space.

The material on $K(\pi, n)$'s is mainly due to Eilenberg and MacLane [13]. The proof of uniqueness of $K(\pi, n)$'s is due to Moore [5, 52]. Our presentation of the results of [13] comparing maps into $K(\pi, n)$'s and cohomology groups follows Douady [7]. The relationship between $K(\pi, n)$'s and cohomology operations was first observed by Serre [58].

The approach to the k-invariants of Postnikov systems via the study of simplicial fibre bundles with fibre $K(\pi, n)$ is new. An obstruction theoretic approach in the semi-simplicial case is taken by Heller [21] and Moore [53].

CHAPTER VI

LOOP GROUPS, ACYCLIC MODELS, AND TWISTED TENSOR PRODUCTS

The main object of this chapter is the construction of the Serre spectral sequence by means of twisted tensor products. The basic tool in this development, which is completed in section 32, is Brown's theorem, which gives a natural equivalence between $C_N(F \times_\tau B)$ and $C_N(B) \otimes_t C_N(F)$ on the category of twisted Cartesian products, where t is a twisting cochain determined by the twisting function τ. In order to define twisting cochains, the theory of cup, Pontryagin, and cap products is developed in section 30. Of course, the definition of these products depends on the Eilenberg-Zilber theorem, and this is proven in section 29. The proofs of both Brown's theorem and the Eilenberg-Zilber theorem rely on the method of acyclic models, which is described in section 28. The models for Brown's theorem are defined in terms of functors which assign to a (reduced) simplicial set K a simplicial group $G(K)$ and a PTCP $G(K) \times_\tau K$ such that $T(E(\tau))$ is contractible. $G(K)$ is called a loop group of K. $G(K)$ and $G(K) \times_\tau K$ are defined in section 26. In section 27, it is shown that G and $\overline{W}$ are adjoint functors, the suspension $E(K)$ of a complex K is defined, and miscellaneous results about the functors G, $\overline{W}$, and E are obtained.

§26. Loop groups

In this section, we will prove the existence of loop groups of reduced complexes, that is, of complexes having just one vertex. We will also give a reinterpretation of the Hurewicz homomorphism.

DEFINITION 26.1: A group complex G is said to be a loop group of the complex K if there exists a PTCP $E(\tau) = G \times_\tau K$ such that $T(E(\tau))$ is a contractible space.

EXAMPLES 26.2: $K(\pi, n)$ is of course a loop group of $K(\pi, n+1)$. By Lemma 23.4, if K is a Kan complex, then $L(K)$ is a loop group of K provided that $L(K)$ admits a structure of simplicial group.

If K is a reduced complex and $K_0 = k_0$, we will let k_n denote $s_0^n k_0$. We now define a loop group of such a complex.

DEFINITION 26.3: Let K be a reduced complex. Define $G_n(K)$ to be the free group generated by the elements of K_{n+1} modulo the relations $s_0 x = e_n$ for $x \in K_n$. $G_n(K)$ is of course a free group. If $x \in K_{n+1}$, let $\tau(x)$ denote the class of x in $G_n(K)$. Define face and degeneracy operators on the generators of $G(K)$ by:

$$\tau(\partial_0 x)\,\partial_0\tau(x) = \tau(\partial_1 x)$$

$$(T) \qquad \partial_i\tau(x) = \tau(\partial_{i+1}x) \quad \text{if } i > 0$$

$$s_i\tau(x) = \tau(s_{i+1}x) \quad \text{if } i \geq 0.$$

The ∂_i and s_i extend uniquely to homomorphisms $G_n(K) \to G_{n-1}(K)$ and $G_n(K) \to G_{n+1}(K)$. $G(K)$ so defined is easily verified to be a group complex, and $\tau : K \to G(K)$ is clearly a twisting function. We let $E(\tau) = G(K) \times_\tau K$.

We must prove that $T(E(\tau))$ is contractible, and it suffices to prove $\pi_1(T(E(\tau))) = 0$ and $\tilde{H}_n(E(\tau)) = 0$, $n \geq 0$.

LEMMA 26.4: $\pi_1(T(E(\tau))) = 0$.

Proof: Recall that $\pi_1(T(E(\tau)))$ can be considered as a group having one generator for each 1-cell not in a maximal tree and one relation for each 2-cell. We regard non-degenerate simplices (g, x) as denoting the corresponding cells. The 1-cells $(s_0 g, x)$, $x \in K_1$ non-degenerate and $g \in G_0(K)$, form a maximal tree. This holds since $\partial_0(s_0 g, x) = (\tau(x)g, k_0)$ and $\partial_1(s_0 g, x) = (g, k_0)$, which implies that any two 0-cells can be connected in one and only one way by 1-cells of the cited form. We must show that every 1-cell (g, x), g non-degenerate, is homotopic to the product of 1-cells in the maximal tree and their inverses (reverses). The 2-cell $(s_1 g, s_0 x)$ shows that (g, x) is homotopic to $(s_0\partial_0 g, x)(g, k_1)$. If $g = \tau(y)^{-1}g'$, the 2-cell $(s_0 g, y)$ shows that $(g, \partial_1 y)$ is homotopic to $(g', \partial_0 y)(s_0\partial_1 g, \partial_2 y)$. If $g = \tau(y)g'$, the 2-cell $(s_0 g', y)$ shows that $(g, \partial_0 y)$ is homotopic to $(g', \partial_1 y)(s_0\partial_1 g', \partial_2 y)^{-1}$. Combining these relations, (g, x) is homotopic to the product of (g', k_1) with 1-cells of the maximal tree or their reverses, where the group theoretic length of g' is strictly less than that of g. Inductively, since $e_1 = s_0 e_0$, the result is proven.

LEMMA 26.5: $\tilde{H}_n(E(\tau)) = 0$, $n \geq 0$.

Proof: Consider $\tilde{C}_n(E(\tau))$, where (e_0, k_0) is taken as base point. For $g \in G_n(K)$ and $x \in K_{n+1}$, define $[g, x] \in \tilde{C}_n(E(\tau))$ by:

(i) $[g, x] = (\tau(x)g, \partial_0 x) - (g, k_n)$.

Observe that $[g, k_{n+1}] = 0$ and define $B = \{[g, x] \mid x \neq k_{n+1}\}$. Suppose for the moment that we know that B is a basis for the free Abelian group $\tilde{C}_n(E(\tau))$, and define $S : \tilde{C}_n(E(\tau)) \to \tilde{C}_{n+1}(E(\tau))$ by

(ii) $S[g, x] = \sum_{i=0}^{n} (-1)^i [s_i g, (s_0)^{i+1}(\partial_1)^i x]$.

Using the easily verified relations

$$\text{(iii)} \quad \partial_i[g,x] = [\partial_i g, \partial_{i+1}x] \text{ and } s_i[g,x] = [s_i g, s_{i+1}x],$$

a simple calculation proves that $dS + Sd = 1$ on $\tilde{C}(E(\tau))$. It remains to prove that B is a basis for $\tilde{C}_n(E(\tau))$. We show first that the subgroup $Z(B)$ of $\tilde{C}_n(E(\tau))$ generated by B is all of $\tilde{C}_n(E(\tau))$. Let (g, x) be an element of the natural basis for $\tilde{C}_n(E(\tau))$. Since $(g,x) = (g,k_n) + [g,s_0x]$, $(g,x) \equiv (g,k_n) \bmod Z(B)$. Since

$$\text{(iv)} \quad (\tau(y)^{-1}g',k_n) = (g,\partial_0 y) - [\tau(y)^{-1}g',y] \text{ and}$$
$$\text{(v)} \quad (\tau(y)g',\partial_0 y) = (g',k_n) + [g,y],$$

and g can be written in one of the forms $g = \tau(y)^{-1}g'$ or $g = \tau(y)g'$, it follows that $(g,x) \equiv (g',k_n) \bmod Z(B)$, where g' is an element of $G_n(K)$ with length strictly less than that of g. Inductively, we find $Z(B) = \tilde{C}_n(E(\tau))$. We must still prove that the elements of B are linearly independent. Let q be an integer and define B_q to be the union of the following sets:

$$\{[g,s_0y] \mid y \neq k_n, \text{ length }(g) \leq q\}$$
$$\text{(vi)} \quad \{[g,x] \mid \text{length }(\tau(x)g) = \text{length }(g) + 1 \leq q\}$$
$$\{[\tau(x)^{-1}g,x] \mid \text{length }(\tau(x)^{-1}g) = \text{length }(g) + 1 \leq q\}$$

Since $B = \bigcup_q B_q$, it suffices to prove that the elements of B_q are linearly independent for each q. $B_0 = \{[e_n, s_0y]\} = \{(e_n,y)\}$, so the result is true for $q = 0$. Now we need only show that the elements of $B_q - B_{q-1}$ are linearly independent mod $Z(B_{q-1})$, $q \geq 1$. But $B_q - B_{q-1}$ mod $Z(B_{q-1})$ is the union of the following sets:

$$\{(g,y) - (g,k_n) \mid \text{length }(g) = q\}$$
$$\text{(vii)} \quad \{(\tau(x)g,\partial_0 x) \mid \text{length }(\tau(x)g) = \text{length }(g) + 1 = q\}$$
$$\{(\tau(x)^{-1}g,k_n) \mid \text{length }(\tau(x)^{-1}g) = \text{length }(g) + 1 = q\}.$$

Clearly these elements are linearly independent, and this completes the proof.

Thus we have proven the following theorem.

THEOREM 26.6: Let K be a reduced complex. Then $T(E(\tau))$ is contractible, hence $G(K)$ is a loop group of K.

COROLLARY 26.7: The homotopy groups $\pi_n(K) = \pi_n(ST(K))$ of a reduced complex K are calculable as $\pi_{n-1}(G(K))$, $n \geq 1$.

REMARKS 26.8: In principle, the corollary defines a procedure for a purely group theoretic calculation of the homotopy groups of any reduced complex. The construction of G can actually be carried out for an arbitrary connected complex. The details of this generalization may be found in Kan [30].

There is an interesting interpretation of the Hurewicz homomorphism in terms of the functor G. Let $A(K) = G(K)/[G(K),G(K)]$. Then the projection $p: G(K) \to A(K)$ can be verified to be a Kan fibration, whose fibre we denote by $F(K)$. We will show that $p_*: \pi_{n-1}(G(K)) \to \pi_{n-1}(A(K))$ is essentially the Hurewicz homomorphism $h: \pi_n(K) \to H_n(K)$.

THEOREM 26.9: $\pi_n(A(K))$ is naturally isomorphic to $H_{n+1}(K)$, $n \geq 0$, and the following diagram commutes

$$\begin{array}{ccc} \pi_{n+1}(K) & \xrightarrow{h} & H_{n+1}(K) \\ \downarrow & & \downarrow \\ \pi_n(G(K)) & \xrightarrow{p_*} & \pi_n(A(K)), \end{array}$$

where the vertical arrows are the natural isomorphisms.

Proof: By Theorem 22.1, if we regard $A(K)$ as a chain complex with differential $\partial = \Sigma(-1)^i\partial_i$, then $\pi_n(A(K)) = H_n(A(K))$. Observe that $A_n(K) = C_{n+1}(K)/s_0C_n(K)$. Thus it suffices to prove that $H_{n+1}(C(K)) = H_{n+1}(C(K)/s_0C(K))$. Now if A is any Abelian group complex, then $\partial s_0 + s_0\partial = s_0\partial_0$ on A_n, $n \geq 1$, hence

$\partial s_0(s_0 x) + s_0\partial(s_0 x) = s_0 x$ for $x \in A_{n-1}$. Therefore $s_0 A$ is stable under ∂ and is null-homotopic. Since the commutativity of the cited diagram follows from the definitions, this proves the result.

REMARKS 26.10: We have the following commutative diagram, where the rows are exact and the vertical arrows are isomorphisms:

$$\begin{array}{ccccccccc} \cdots \to & \pi_{n+1}(A(K) \times_\tau K) & \to & \pi_{n+1}(K) & \to & H_{n+1}(K) & \to & \pi_n(A(K) \times_\tau K) & \to \cdots \\ & \downarrow & & \downarrow & & \downarrow & & \downarrow & \\ \cdots \to & \pi_n(F(K)) & \to & \pi_{n+1}(K) & \to & H_{n+1}(K) & \to & \pi_{n-1}(F(K)) & \to \cdots \end{array}$$

The first exact sequence arises from $p\colon G(K) \to A(K)$, the second from the PTCP $A(K) \times_\tau K$. The outer vertical arrows result from the fibration $G(K) \times_\tau K \to A(K) \times_\tau K$ with fibre $F(K)$.

§27. The functors G, $\overline{W}$, and E

Here we prove that G and $\overline{W}$ are adjoint functors. From this will follow the essential uniqueness of loop groups. We then define the suspension $E(K)$ of a complex K and make some remarks about the functor E.

THEOREM 27.1: Let $\mathcal{G}$ denote the category of simplicial groups, $\mathcal{S}$ that of reduced complexes. Then the functor $G\colon \mathcal{S} \to \mathcal{G}$ is an adjoint of $\overline{W}\colon \mathcal{G} \to \mathcal{S}$.

Proof: Define $\psi\colon \mathrm{Hom}_{\mathcal{G}}(G(K),A) \to \mathrm{Hom}_{\mathcal{S}}(K,\overline{W}(A))$ as follows. Given a simplicial homomorphism $\gamma\colon G(K) \to A$, there exists a unique γ-equivariant map $\theta\colon G(K) \times_\tau K \to A$ such that $\theta(e_q \times K_q) \subset e_q \times \overline{W}_q(A)$, $q \geq 0$ (by Theorem 21.7). Define $\psi(\gamma)\colon K \to \overline{W}(A)$ by $p\theta = \psi(\gamma)p$. As in (1) of Lemma 21.9, $\psi(\gamma)$ is given explicitly by:

(1) $\psi(\gamma)(x) = [\gamma\tau(x), \gamma\tau(\partial_0 x), \ldots, \gamma\tau(\partial_0^{n-1} x)]$, $x \in K_n$.

Define $\phi\colon \mathrm{Hom}_{\mathcal{S}}(K,\overline{W}(A)) \to \mathrm{Hom}_{\mathcal{G}}(G(K),A)$ by

(2) $\phi(f)\,(\tau(x)) = \tau(A)\,(f(x)),\ \ x \in K_n$.

This means that if $f(x) = [a_{n-1}, \ldots, a_0]$, then $\phi(f)\,(\tau(x)) = a_{n-1}$. $\phi(f)$ is extended to $G(K)$ by the requirement that it be a simplicial homomorphism. It is easily verified that $\phi \circ \psi$ and $\psi \circ \phi$ are the respective identities.

COROLLARY 27.2: There is a natural one-to-one correspondence between twisting functions $K \to A$ and simplicial homomorphisms $G(K) \to A$.

Proof: Each twisting function $\overline{\tau}\colon K \to A$ is induced by a unique simplicial map $f(\overline{\tau})\colon K \to \overline{W}(A)$ by the rule $\overline{\tau} = \tau(A) \circ f(\overline{\tau})$. Notice that

(3) $\phi(f(\overline{\tau}))\,(\tau(x)) = \tau(A)f(\overline{\tau})\,(x) = \overline{\tau}(x),\ \ x \in K_n$.

COROLLARY 27.3: Any PTCP with base K is determined by a unique simplicial homomorphism of $G(K)$ into the group of the PTCP.

COROLLARY 27.4: Let A be a loop complex for K. Then there exists a simplicial homomorphism $G(K) \to A$ which is a homotopy equivalence.

Proof: There exists a twisting function $\overline{\tau}\colon K \to A$ such that $T(A \times_{\overline{\tau}} K)$ is contractible. If $f(\overline{\tau})\colon K \to \overline{W}(A)$ induces $A \times_{\overline{\tau}} K$, then clearly $f(\overline{\tau})$ is a weak homotopy equivalence. Consider $\phi(f(\overline{\tau}))\colon G(K) \to A$. By definition of ψ, the $\phi(f(\overline{\tau}))$-equivariant map $G(K) \times_{\tau} K \to W(A)$ of Theorem 21.7 covers $(\psi \circ \phi)(f(\overline{\tau})) = f(\overline{\tau})\colon K \to \overline{W}(A)$. Therefore $\phi(f(\overline{\tau}))$ is a weak homotopy equivalence, hence a homotopy equivalence.

REMARKS 27.5: Free groups play a role in the category $\mathcal{G}$ similar to that played by Kan complexes in the category $\mathcal{S}$. To see this, we note first that the natural concept of homotopy in $\mathcal{G}$ is that of loop homotopy: Two simplicial homomorphisms $f,g\colon G \to H$ are loop homotopic if there exists a homotopy $F\colon f \simeq g$ such that $F(xy,v) = F(x,v)F(y,v)$ for all $x,y \in G_n$ and $v \in I_n$. It can be shown, by appropriately redefining H^G in the category $\mathcal{G}$ and proving that H^G is a Kan complex if G is a free group, that loop homotopy is an equivalence relation on homomorphisms defined on free groups. Similarly, a weak homotopy equivalence of free groups is a loop homotopy equivalence. Further, ϕ and ψ of Theorem 27.1 induce one-to-one correspondences between the homotopy classes of simplicial maps $K \to \overline{W}(A)$ and the loop homotopy classes of simplicial homomorphisms $G(K) \to A$. Let $\Phi\colon G\overline{W} \to 1_{\mathcal{G}}$ and $\Psi\colon 1_{\mathcal{S}} \to \overline{W}G$ correspond to ϕ and ψ (as in Lemma 15.2) so that $\overline{W}\Phi \circ \Psi\overline{W} : \overline{W} \to \overline{W}$ and $\Phi G \circ G\Psi : G \to G$ are the respective identity natural transformations. In analogy with Theorem 16.6, we can prove that $\Psi(K) : K \to \overline{W}G(K)$ is a weak homotopy equivalence (and is a homotopy equivalence if K is a Kan complex) and that $\Phi(A) : G\overline{W}(A) \to A$ is a weak homotopy equivalence (and is a loop homotopy equivalence if A is a free group complex). Finally, $G(F) : G(f) \simeq G(g)$ is a loop homotopy in $\mathcal{G}$ if $F\colon f \simeq g$ in $\mathcal{S}$, and $\overline{W}(F) : \overline{W}(f) \simeq \overline{W}(g)$ in $\mathcal{S}$ if $F : f \simeq g$ is a loop homotopy in $\mathcal{G}$. Proofs of these results may be found in Kan [31].

Now we define the suspension of a complex.

DEFINITION 27.6: Let K be a complex with base point k_0. Define the suspension $E(K)$ as follows. $E_0(K) = b_0$; $E_n(K)$ consists of all symbols (i,x), $i \geq 1$, $x \in K_{n-i}$, modulo the identification $(i,k_n) = s_0^{n+i}b_0 = b_{n+i}$, where $k_n = s_0^n k_0$. Define face and degeneracy operators on $E(K)$ by:

(i) $s_0^i(j,x) = (i+j,x)$

(ii) $s_{i+1}(1,x) = (1,s_i x)$

(iii) $\partial_0(1,x) = b_n, \quad x \in K_n$

(iv) $\partial_1(1,x) = b_0, \quad x \in K_0$

(v) $\partial_{i+1}(1,x) = (1,\partial_i x), \quad x \in K_n, \quad n > 0.$

$s_i(j,x)$ for $i \geq 1$, $j \geq 2$ and $\partial_i(j,x)$ for $i \geq 0$, $j \geq 2$ are determined by the operations above and the requirement that $E(K)$ be a complex. We note that $(i+1,x)$ is often denoted by $s_0^i Ex$ in the literature.

Recall that if (X,x_0) is a topological pair, then the (reduced) suspension $E(X)$ of X is the identification space obtained from $X \times I$ by collapsing $(X \times \dot{I}) \cup (x_0 \times I)$ to a point.

LEMMA 27.7: $TE(K)$ is canonically homeomorphic to $ET(K)$.

Proof: Define $f\colon ET(K) \to TE(K)$ by the formula:

(i) $f(|x,\delta|,1-t) = |(1,x),\delta(t)|$, where $x \in K_n$ and where if $\delta = (t_0,\ldots,t_n)$, then $\delta(t) = (1-t,tt_0,\ldots,tt_n)$.

f is the required homeomorphism.

Writing $\tau(1,x) = x$, we find that $GE(K)$ is just the free simplicial group generated by K, subject only to the relations $k_n = e_n$. It follows that $AE(K) = \tilde{C}(K)$, so that $\pi_n(AE(K)) = \tilde{H}_n(K)$. The inclusion map $K \to GE(K)$ induces $\sigma\colon \pi_n(K) \to \pi_n(GE(K)) \cong \pi_{n+1}(EK)$. σ is called the suspension map of the homotopy groups. Using Theorem 26.9, we obtain the commutative diagram:

$$\begin{array}{ccccc} & & \sigma & & \\ \pi_n(K) & \longrightarrow & \pi_n(GE(K)) & \xrightarrow{\cong} & \pi_{n+1}(E(K)) \\ \downarrow h & & \downarrow p_* & & \downarrow h \\ \tilde{H}_n(K) & \xrightarrow{\cong} & \pi_n(AE(K)) & \xrightarrow{\cong} & H_{n+1}(E(K)). \end{array}$$

The pair $(GE(K),K)$ gives rise to the exact sequence:

$$\ldots \to \pi_{n+1}(GE(K),K) \to \pi_n(K) \xrightarrow{\sigma} \pi_{n+1}(E(K)) \to \pi_n(GE(K),K) \to \ldots$$

REMARKS 27.8: The construction GE could be modified by using free monoids instead of free groups. We denote the resulting free monoid complex by $G^+E(K)$. The interest in $G^+E(K)$ arises from the fact that if K is a countable complex, then $TG^+E(K)$ is canonically homeomorphic to the reduced product (James [23]) of $T(K)$. However, a homological argument shows that the inclusion of $G^+E(K)$ in $GE(K)$ induces an isomorphism on homotopy groups when K is connected.

REMARKS 27.9: Let $\mathcal{S}$ denote the category of simplicial sets, $\mathcal{S}^+$ that of reduced complexes. One would like $G: \mathcal{S}^+ \to \mathcal{S}$ and $E: \mathcal{S} \to \mathcal{S}^+$ to be adjoint functors, since in the topological case we have $\mathrm{Hom}_{\mathcal{T}}(X,\Omega Y) \cong \mathrm{Hom}_{\mathcal{T}}(E(X),Y)$, where $\Omega(Y)$ denotes the loop space of Y. This conclusion is false: It is easy to construct a natural transformation $\psi\colon \mathrm{Hom}_{\mathcal{S}^+}(E(K),K') \to \mathrm{Hom}_{\mathcal{S}}(K,G(K'))$, but ψ has no inverse. If G is replaced by L of Definition 23.3 and $\mathcal{S}$ is taken to be the category of Kan complexes, then ψ is still defined and now does have an inverse. Thus the functor L is more closely analogous to Ω than is G. However, $L(K)$ is only defined if K is a Kan complex, and $L(K)$ may admit no structure of group complex.

§28. Acyclic models

We develop here a general procedure for the construction of

chain maps and chain homotopies. The method applies when a category has certain distinguished objects, called models, which are acyclic with respect to a homology theory defined in terms of one functor from the given category to the category of chain complexes and which, in a suitable sense, represent another such functor.

Let $\mathcal{S}$ denote a fixed category. Let $\mathcal{M}$ denote a set of objects of $\mathcal{S}$, called the models. $\mathcal{C}$ will denote the category of chain complexes, $\mathcal{B}$ that of Abelian groups. Let $A: \mathcal{S} \to \mathcal{B}$ be a covariant functor. We must define the concept of representability of A. Assume (or arrange by use of the free Abelian group functor) that $\mathcal{S}$ is an additive category and A is an additive functor. Define $\tilde{A}(K) = \oplus_{M \epsilon \mathcal{M}} (\mathrm{Hom}_{\mathcal{S}}(M,K) \otimes A(M))$, K an object of $\mathcal{S}$. For maps $\alpha: K \to L$ and $\mu: M \to K$ in $\mathcal{S}$ and $a \epsilon A(M)$, define $\tilde{A}(\alpha)(\mu,a) = (\alpha \circ \mu, a)$.

Then $\tilde{A}: \mathcal{S} \to \mathcal{B}$ is an additive functor. Define a natural transformation of functors $\lambda: \tilde{A} \to A$ by $\lambda(K)(\mu,a) = A(\mu)(a)$, $\mu: M \to K$, $a \epsilon A(M)$.

DEFINITION 28.1: $A: \mathcal{S} \to \mathcal{B}$ is said to be representable if there exists a natural transformation of functors $\xi: A \to \tilde{A}$ such that $\lambda\xi: A \to A$ is the identity natural transformation. ξ is then said to be a representation of A.

THEOREM 28.2: Let A and B be additive functors $\mathcal{S} \to \mathcal{C}$ and let $n \geq 0$ be an integer. Suppose given natural transformations $f_i: A_i \to B_i$ such that $d_i f_i = f_{i-1} d_i$, $0 \leq i \leq n$ (where $A_i, B_i: \mathcal{S} \to \mathcal{B}$ are determined by $A_i(K)$, $B_i(K)$, K an object of $\mathcal{S}$). If each $A_q: \mathcal{S} \to \mathcal{B}$ is representable for $q > n$ and each $H_q(B(M)) = 0$ for $q \geq n$ and $M \epsilon \mathcal{M}$, then there exists a natural transformation $f: A \to B$ such that $f \mid A_i = f_i$, $0 \leq i \leq n$.

Proof: We construct $f_q: A_q \to B_q$ by induction on q, starting with the given f_i, $i \leq n$. Suppose f_{q-1} has been defined on

A_{q-1}, $q \geq n+1$. If $a \in A_q(M)$, $M \in \mathfrak{M}$, then $f_{q-1}d_q(a) \in Z_{q-1}B(M)$ because by the induction hypothesis we have

$$d_{q-1}f_{q-1}d_q = f_{q-2}d_{q-1}d_q = 0.$$

Since $H_{q-1}(B(M)) = 0$, we may choose $b \in B_q(M)$ such that $d_q(b) = f_{q-1}d_q(a)$. Define a natural transformation $\mathfrak{z}: \tilde{A}_q \to B_q$ by $\mathfrak{z}(K)(\mu,a) = B_q(\mu)(b)$, $\mu: M \to K$. Then we find:

$$\begin{aligned} d_q\mathfrak{z}(K)(\mu,a) &= d_qB_q(\mu)(b) = B_{q-1}(\mu)d_q(b) = B_{q-1}(\mu)f_{q-1}d_q(a) \\ &= f_{q-1}A_{q-1}(\mu)d_q(a) = f_{q-1}d_qA_q(\mu)(a) = f_{q-1}d_q\lambda(K)(\mu,a). \end{aligned}$$

Thus $d_q\mathfrak{z} = f_{q-1}d_q\lambda$. Now let $\xi: A_q \to \tilde{A}_q$ be a representation of A_q and define $f_q = \mathfrak{z}\xi: A_q \to B_q$. Then we have

$$d_qf_q = d_q\mathfrak{z}\xi = f_{q-1}d_q\lambda\xi = f_{q-1}d_q,$$

as desired.

THEOREM 28.3: Let A and B be **additive** functors $\mathcal{S} \to \mathcal{C}$ and let $n \geq 0$ be an integer. Let f and g be natural transformations $A \to B$ and suppose given natural transformations $s_i: A_i \to B_{i+1}$ such that $d_{i+1}s_i + s_{i-1}d_i = f_i - g_i$, $0 \leq i \leq n-1$. If each $A_q: \mathcal{S} \to \mathcal{B}$ is representable for $q \geq n$ and each $H_q(B(M)) = 0$ for $q \geq n$ and $M \in \mathfrak{M}$, then there exists a natural chain homotopy $s: f \simeq g$ such that $s \mid A_i = s_i$, $0 \leq i \leq n-1$.

Proof: The argument is similar to that above. We construct $s_q: A_q \to B_{q+1}$ by induction on q, starting with the given s_i, $i \leq n-1$, or with $s_{-1} = 0$ if $n = 0$. Suppose s_{q-1} has been defined on A_{q-1}, $q \geq n$. If $a \in A_q(M)$, $M \in \mathfrak{M}$, then we find that $(g_q - f_q - s_{q-1}d_q)(a) \in Z_qB(M)$ by use of the induction hypothesis. We then choose $c \in B_{q+1}(M)$ such that $d_{q+1}(c) = (g_q - f_q - s_{q-1}d_q)(a)$ and define a natural transformation $\eta: \tilde{A}_q \to B_{q+1}$ by

$$\eta(K)(\mu,a) = B_{q+1}(\mu)(c), \quad \mu: M \to K.$$

An easy calculation proves that $d_{q+1}\eta = (g_q - f_q - s_{q-1}d_q)\lambda$. Letting $\xi: A_q \to \tilde{A}_q$ be a representation of A_q, we define

$$s_q = \eta\xi: A_q \to B_{q+1}$$

and then $d_{q+1}s_q = g_q - f_q - s_{q-1}d_q$, as desired.

REMARKS 28.4: In the applications, one often encounters the situation where $B(M)$ is an augmented chain complex satisfying $H_q(B(M)) = 0$ for $q > 0$ and $\varepsilon: H_0(B(M)) \xrightarrow{\cong} Z$. In this case, we say that M is acyclic with respect to B. Here, if $n = 0$ in Theorem 28.2, we must assume that $f_0\partial_1(A(M)) \subset \operatorname{Ker} \varepsilon$ for all $M \in \mathfrak{M}$, while if $n = 0$ in Theorem 28.3 we must assume that

$$(f_0 - g_0)\,(A(M)) \subset \operatorname{Ker}\, \varepsilon.$$

The proofs then go through with the obvious trivial modifications.

REMARKS 28.5: If each model M has a canonical contracting homotopy with respect to B, then there are canonical choices for the chains b and c in the preceding proofs. Therefore explicit formulas for the constructed chain maps and homotopies are then obtainable.

§29. The Eilenberg-Zilber theorem

Let $\mathfrak{A}$ denote the category of simplicial Abelian groups. If K and L are objects of $\mathfrak{A}$, it is necessary to modify the definition of $K \times L$ by letting $(K \times L)_n = K_n \otimes L_n$ in order for $K \times L$ to be an object of $\mathfrak{A}$. Now we have two functors A and B: $\mathfrak{A} \times \mathfrak{A} \to \mathcal{C}$, defined as follows. $A(K, L)$ is $A(K \times L)$, that is, $K \times L$ regarded as a chain complex, with $d_n = \Sigma_{i=0}^{n} (-1)^i \partial_i$. $B(K,L)$ is $A(K) \otimes A(L)$, so that $B_n(K,L) = \Sigma_{i+j=n} A_i(K) \otimes A_j(L)$ and $d = d \otimes 1 + 1 \otimes d$ (with the sign convention: $(1 \otimes d)(x \otimes y) = (-1)^{\deg x} x \otimes d(y)$). We will obtain a natural chain homotopy equivalence of functors $f: A \to B$

with inverse $g: B \to A$ by use of the method of acyclic models.

Since $A_0(K,L) = K_0 \otimes L_0 = B_0(K,L)$, we define $f_0: A_0 \to B_0$ and $g_0: B_0 \to A_0$ to be the identity natural transformations. We will obtain models in $\mathfrak{A} \times \mathfrak{A}$ which represent both A and B and are acyclic with respect to both A and B. Thus let M^q denote the free simplicial Abelian group generated by $\Delta[q]$ and define the models $\mathfrak{M}$ in $\mathfrak{A} \times \mathfrak{A}$ to be the pairs (M^p, M^q).

LEMMA 29.1: Both A_n and B_n are representable, $n \geq 0$.

Proof: Define $\xi: A_n \to \tilde{A}_n$ by

(1) $\xi(K,L)(k_n \otimes \ell_n) = (\bar{k}_n \times \bar{\ell}_n, \Delta_n \otimes \Delta_n)$.

Then

$$\lambda(K,L)\xi(K,L)(k_n \otimes \ell_n) = A(\bar{k}_n \times \bar{\ell}_n)(\Delta_n \otimes \Delta_n) = \bar{k}_n(\Delta_n) \otimes \bar{\ell}_n(\Delta_n) = k_n \otimes \ell_n.$$

Similarly, define $\xi: B_n \to \tilde{B}_n$ by

(2) $\xi(K,L)(k_p \otimes \ell_q) = (\bar{k}_p \times \bar{\ell}_q, \Delta_p \otimes \Delta_q)$, $p + q = n$.

Again we find $\lambda\xi = 1$, as desired.

LEMMA 29.2: The models are acyclic with respect to both A and B.

Proof: M^p is augmented with augmentation $\varepsilon: M^p \to Z$ defined by $\varepsilon(x) = 0$ if $x \in M^p_n$, $n > 0$, and $\varepsilon(x) = 1$ if $x \in \Delta[p]_0$. Define $\varepsilon: A(M^p, M^q) \to Z$ by $\varepsilon(x \otimes y) = \varepsilon(x)\varepsilon(y)$, and similarly for $B(M^p, M^q)$. We must prove that $H_n(A(M^p, M^q)) = 0$, $n > 0$, and $\varepsilon: H_0(A(M^p, M^q)) \xrightarrow{\cong} Z$, and similarly for B. Define $s: M^p_n \to M^p_{n+1}$ by $s(a_0, \ldots, a_n) = (0, a_0, \ldots, a_n)$, $0 \leq a_0 \leq \ldots \leq a_n \leq p$. Then

(1) $\partial_0 s = 1$; $\partial_1 s = (0)$ on $\Delta[p]_0$; $\partial_{i+1} s = s\partial_i$ on $\Delta[p]_n$, $n > 0$.

Therefore $ds + sd = 1 - \sigma\varepsilon$, where $\sigma: Z \to M^p$ is defined by

$\sigma(1) = (0)$. Now define $s\colon A_n(M^p,M^q) \to A_{n+1}(M^p,M^q)$ by

(2) $s(x \otimes y) = s(x) \otimes s(y),\quad x \in M^p_n,\quad y \in M^q_n.$

Then $ds + sd = 1 \otimes 1 - \sigma\mathcal{E} \otimes \sigma\mathcal{E}$ on $A(M^p,M^q)$. Similarly define $s\colon B_n(M^p,M^q) \to B_{n+1}(M^p,M^q)$ by

(3) $s(x \otimes y) = s(x) \otimes y + (-1)^v \sigma\mathcal{E}(x) \otimes s(y),\quad x \in M^p_u,\quad y \in M^q_v,$

$u + v = n.$

Then $ds + sd = 1 \otimes 1 - \sigma\mathcal{E} \otimes \sigma\mathcal{E}$ on $B(M^p,M^q)$. This proves the result.

Now the following theorem, due to Eilenberg and Zilber [16], follows immediately from Theorem 28.2 (with $n = 0$) and Theorem 28.3 (with $n = 1$ and $s_0 = 0$), coupled with Remarks 28.4.

THEOREM 29.3: Let $A, B\colon \mathcal{A} \times \mathcal{A} \to \mathcal{C}$ be defined by

$$A(K,L) = A(K \times L)$$

and $B(K,L) = A(K) \otimes A(L)$. Then there exist natural transformations $f\colon A \to B$ and $g\colon B \to A$ lying over the respective identity transformations $f_0\colon A_0 \to B_0$ and $g_0\colon B_0 \to A_0$. Any two such f are naturally chain homotopic, as are any two such g. Further, for any such f and g, $f \circ g\colon B \to B$ and $g \circ f\colon A \to A$ are naturally chain homotopic to the respective identity transformations.

COROLLARY 29.4: Let $A_N, B_N\colon \mathcal{A} \times \mathcal{A} \to \mathcal{C}$ be defined by $A_N(K,L) = A_N(K \times L)$ and $B_N(K,L) = A_N(K) \otimes A_N(L)$. Then the theorem remains true if we replace A, B by A_N, B_N.

Proof: The models of $\mathcal{A} \times \mathcal{A}$ are acyclic with respect to A_N and B_N by Corollary 22.3. It is easy to check directly that A_N and B_N are representable, but since it follows from Theorem 22.2 and Corollary 22.3 that $A_N(K,L)$ and $B_N(K,L)$ are direct summands of $A(K,L)$ and $B(K,L)$, this result is a consequence of the following lemma, which we will have further need of later.

LEMMA 29.5: Let $A, A': \mathcal{S} \to \mathcal{C}$ be additive functors. Let $\alpha: A \to A'$ and $\beta: A' \to A$ be natural transformations such that $\beta \circ \alpha = 1: A \to A$. Then if A' is representable, so is A.

Proof: The following diagram is commutative:

$$\begin{array}{ccccc} \tilde{A} & \xrightarrow{\tilde{\alpha}} & \tilde{A}' & \xrightarrow{\tilde{\beta}} & \tilde{A} \\ \downarrow \lambda & & \downarrow \lambda' & & \downarrow \lambda \\ A & \xrightarrow{\alpha} & A' & \xrightarrow{\beta} & A . \end{array}$$

Here $\tilde{\alpha}(K)(\mu,a) = (\mu,\alpha(M)(a))$, $\mu: K \to M$, $a \in A(M)$, and $\tilde{\beta}(K)(\mu,a') = (\mu,\beta(M)(a'))$, $a' \in A'(M)$. Let $\xi': A' \to \tilde{A}'$ be a representation of A' and define $\xi: A \to \tilde{A}$ by $\xi = \tilde{\beta}\xi'\alpha$. Then: $\lambda\xi = \lambda\tilde{\beta}\xi'\alpha = \beta\lambda'\xi'\alpha = \beta\alpha = 1$, so ξ is a representation of A.

COROLLARY 29.6: Let K and L be simplicial sets. Then there are natural chain homotopy equivalences $f: C(K \times L) \to C(K) \otimes C(L)$ and $g: C(K) \otimes C(L) \to C(K \times L)$, and similarly with C replaced by C_N.

Proof: We need only observe that $C(K \times L) = A(F(K),F(L))$ and $C(K) \otimes C(L) = B(F(K),F(L))$, where $F(K)$ and $F(L)$ are the free simplicial Abelian groups generated by K and L.

As stated in Remarks 28.5, the contracting homotopies on the models determine explicit choices for f and g. It is simpler in this case to construct explicit transformations ad hoc. For convenience, if K is a simplicial set and $x \in K_n$, we let $\tilde{\partial}x = \partial_n x$ and in general $\tilde{\partial}^{n-i}x = \partial_{i+1} \cdots \partial_n x$.

DEFINITIONS 29.7: Let K and L be simplicial Abelian groups. Define $f: A(K \times L) \to A(K) \otimes A(L)$ by

(i) $f(k_n \otimes \ell_n) = \sum_{i=0}^{n} \tilde{\partial}^{n-i} k_n \otimes \partial_0^i \ell_n$

f is called the Alexander-Whitney map.

Define $g\colon A(K) \otimes A(L) \to A(K \times L)$ by

(ii) $$g(k_p \otimes \ell_q) = \sum_{(\mu,\nu)} (-1)^{\sigma(\mu)}(s_{\nu_q} \cdots s_{\nu_1} k_p \otimes s_{\mu_p} \cdots s_{\mu_1} \ell_q),$$

where the sum is taken over all (p,q)-shuffles (μ,ν) (Definition 6.5) and where $\sigma(\mu) = \Sigma_{i=1}^{p} [\mu_i - (i-1)]$ is the signature of the corresponding permutation.

We will call g the Eilenberg-MacLane map, since these authors introduced and studied it in [13].

PROPOSITION 29.8: The Alexander-Whitney map f is a natural transformation $A(K \times L) \to A(K) \otimes A(L)$ lying over the identity $A_0(K \times L) \to A_0(K) \otimes A_0(L)$. It is associative in the sense that the following diagram commutes:

$$\begin{array}{ccccc} & & A(K) \otimes A(L \times M) & & \\ & \overset{f}{\nearrow} & & \searrow^{1 \otimes f} & \\ A(K \times L \times M) & & & & A(K) \otimes A(L) \otimes A(M). \\ & \underset{f}{\searrow} & & \nearrow_{f \otimes 1} & \\ & & A(K \times L) \otimes A(M) & & \end{array}$$

f induces a natural transformation $f_N\colon A_N(K \times L) \to A_N(K) \otimes A_N(L)$.

Proof: Clearly f_0 is the identity and f is natural. To prove that $df = fd$ and $(1 \otimes f)f = (f \otimes 1)f$, we need only demonstrate these equations on $\Delta_n \otimes \Delta_n$ and on $\Delta_n \otimes \Delta_n \otimes \Delta_n$. Here these results follow from explicit calculations, which will be omitted. Next, we have an exact sequence (with the obvious maps)

(a) $D(K) \otimes A(L) \oplus A(K) \otimes D(L) \to A(K) \otimes A(L) \to A_N(K) \otimes A_N(L)$.

Now if $x \otimes y \in D(K \times L)$, say $x \otimes y = s_k x' \otimes s_k y'$, then if $i \leq k$, $\partial_0^i s_k y'$ is degenerate and if $i > k$, $\tilde{\partial}^{n-i} s_k x'$ is degenerate. Thus one factor in each term of $f(x \otimes y)$ is degenerate, and therefore, by (a), f does induce f_N.

PROPOSITION 29.9: The Eilenberg-MacLane map g is a natural transformation $A(K) \otimes A(L) \to A(K \times L)$ lying over the identity $A_0(K) \otimes A_0(L) \to A_0(K \times L)$. It is associative in the sense that the following diagram commutes:

$$\begin{array}{ccccc} & & A(K \times L) \otimes A(M) & & \\ & \overset{g \otimes 1}{\nearrow} & & \overset{g}{\searrow} & \\ A(K) \otimes A(L) \otimes A(M) & & & & A(K \times L \times M) \\ & \underset{1 \otimes g}{\searrow} & & \underset{g}{\nearrow} & \\ & & A(K) \otimes A(L \times M) & & \end{array}$$

g induces a natural transformation $g_N: A_N(K) \otimes A_N(L) \to A_N(K \times L)$.

Proof: Clearly g_0 is the identity and g is natural. To prove that $dg = gd$ and $g(g \otimes 1) = g(1 \otimes g)$, it suffices to obtain these results on $\Delta_p \otimes \Delta_q$ and on $\Delta_p \otimes \Delta_q \otimes \Delta_r$. Here these equations follow from lengthy computations, which will be omitted. Finally, if either $x \in K_p$ or $y \in L_q$ is degenerate, then so is each term of $g(x \otimes y)$, and this implies that g induces g_N.

COROLLARY 29.10: $g \circ f \simeq 1$, $f \circ g \simeq 1$, $g_N \circ f_N \simeq 1$, and $f_N \circ g_N = 1$.

Proof: The existence of the cited chain homotopies follows from Theorem 29.3 and Corollary 29.4. The fact that $f_N \circ g_N = 1$ (with no chain homotopy required) is verified by explicit computation.

Finally, f and g are homotopy commutative in the sense of the following lemma.

LEMMA 29.11: Let K and L be simplicial Abelian groups. Define $t: K \times L \to L \times K$ by $t(x \otimes y) = y \otimes x$. Then the following diagrams are chain homotopy commutative:

$$\begin{array}{ccc} A(K \times L) & \xrightarrow{t} & A(L \times K) \\ \downarrow f & & \downarrow f \\ A(K) \otimes A(L) & \xrightarrow{T} & A(L) \otimes A(K) \end{array} \quad \text{and} \quad \begin{array}{ccc} A(K) \otimes A(L) & \xrightarrow{T} & A(L) \otimes A(K) \\ \downarrow g & & \downarrow g \\ A(K \times L) & \xrightarrow{t} & A(L \times K) \end{array},$$

where $T(x \otimes y) = (-1)^{\deg x \deg y} y \otimes x$.

Proof: We note first that the sign in the definition of T makes it a chain map. Since $ft = Tf$ on $A_0(K \times L)$ and $gT = tg$ on $A_0(K) \otimes A_0(L)$, the result follows from Theorem 28.3 (with $n = 1$ and $s_0 = 0$), using the same models as those used in Theorem 29.3.

§30. Cup, Pontryagin, and cap products; twisting cochains

Here we define cup products, Pontryagin products, cap products, and twisting cochains. Twisting cochains define twisted tensor products, which are related to tensor products in a way analogous to the way twisted Cartesian products are related to Cartesian products.

We first define algebras, coalgebras, and Hopf algebras.

DEFINITIONS 30.1: Let X be a graded Λ-module, Λ a commutative ring. X is said to be a Λ-algebra if there exist morphisms of Λ-modules $\phi: X \otimes_\Lambda X \to X$ (the product), $\eta: \Lambda \to X$ (the unit; $\eta(1)$ is the identity), and $\varepsilon: X \to \Lambda$ (the augmentation) such that the following diagrams commute (all tensor products are over Λ in this definition):

$$\begin{array}{ccccc} & & X \otimes X & & \\ & \overset{\phi \otimes 1}{\nearrow} & & \overset{\phi}{\searrow} & \\ X \otimes X \otimes X & & & & X, \\ & \underset{1 \otimes \phi}{\searrow} & & \underset{\phi}{\nearrow} & \\ & & X \otimes X & & \end{array} \qquad \begin{array}{ccccccc} & \overset{\cong}{\nearrow} & \Lambda \otimes X & \xrightarrow{\eta \otimes 1} & X \otimes X & \overset{\phi}{\searrow} & \\ X & & & \xrightarrow{\quad 1 \quad} & & & X, \\ & \underset{\cong}{\searrow} & X \otimes \Lambda & \xrightarrow{1 \otimes \eta} & X \otimes X & \underset{\phi}{\nearrow} & \end{array}$$

$$\begin{array}{ccc} X \otimes X & \xrightarrow{\phi} & X \\ \downarrow{\scriptstyle \varepsilon \otimes \varepsilon} & & \downarrow{\scriptstyle \varepsilon} \\ \Lambda \otimes \Lambda & \xrightarrow{\phi} & \Lambda \end{array} .$$

X is said to be commutative if the following diagram commutes:

$$\begin{array}{ccc} X \otimes X & \xrightarrow{\phi} & X \\ \downarrow T & \nearrow \phi & \\ X \otimes X & & \end{array} \qquad T(x \otimes y) = (-1)^{\deg x \deg y} y \otimes x .$$

Reversing all the arrows, we obtain the definition of a Λ-coalgebra (with coproduct, counit, and coaugmentation), and of cocommutativity of a coalgebra. If X is an algebra with product ϕ, $X \otimes X$ is an algebra with product $(\phi \otimes \phi)(1 \otimes T \otimes 1)$. If X is a coalgebra with coproduct ψ, $X \otimes X$ is a coalgebra with coproduct

$$(1 \otimes T \otimes 1)(\psi \otimes \psi).$$

If X is both an algebra and a coalgebra, and if the unit is the coaugmentation, the augmentation is the counit, and the product is a morphism of coalgebras, then X is said to be a Hopf algebra. The last condition is that $\psi\phi = (\phi \otimes \phi)(1 \otimes T \otimes 1)(\psi \otimes \psi)$ on $X \otimes X$. If X is both an algebra and a differential Λ-module and if $d\phi = \phi d$ ($d = d \otimes 1 + 1 \otimes d$ on $X \otimes X$), then X is said to be a differential algebra. If, further, Y is both a (left) X-module and a differential Λ-module, then Y is said to be a differential X-module if $d\sigma = \sigma d$, where $\sigma\colon X \otimes Y \to Y$ defines the X-module structure on Y. Differential coalgebras and comodules are defined similarly.

In what follows, if the base ring Λ is not mentioned it is assumed to be Z, and $\otimes$ and Hom without subscripts are taken over Z. We will here work with unnormalized chain and cochain complexes, but by Corollary 22.3 (and Lemma 29.5 in acyclic model arguments) we could equally well use normalized complexes.

Let (B, b_0) be a simplicial pair. Define $\Delta\colon B \to B \times B$ by $\Delta(b) = (b, b)$. Let f be the Alexander-Whitney map and define

(1) $D = f \circ C(\Delta)\colon C(B) \to C(B) \otimes C(B)$.

Define $\varepsilon\colon C(B) \to Z$ by $\varepsilon(b) = 1$ if $b \in B_0$ and $\varepsilon(b) = 0$ if $b \in B_n$, $n > 0$. Define $\eta\colon Z \to C(B)$ by $\eta(1) = b_0$. Since $(\Delta \times 1)\Delta = (1 \times \Delta)\Delta$

and $(f \otimes 1)f = (1 \otimes f)f$, we find $(D \otimes 1)D = (1 \otimes D)D$. Clearly $(\varepsilon \otimes 1)D = 1 \otimes 1 = (1 \otimes \varepsilon)D$ and $D\eta = (\eta \otimes \eta)D$. Finally, $dD = Dd$. Therefore $C(B)$ is a differential coalgebra with coproduct D, counit ε, and coaugmentation η. (The counit is, of course, usually called the augmentation).

Now let X be a differential Λ-module with product π, Λ a commutative ring. Consider $\operatorname{Hom}(C(B),X) = C^*(B;X)$. $h \in C^n(B;X)$ is a sequence of homomorphisms $h_q\colon C_q(B) \to X_{q-n}$. The Λ-module structure on X induces such a structure on $C^*(B;X)$. Define the differential δ on $C^*(B;X)$ by

(2) $\delta(h)(b) = dh(b) + (-1)^{\deg h+1}hd(b)$.

Then define the cup product $C^*(B;X) \otimes_\Lambda C^*(B;X) \to C^*(B;X)$ by

(3) $(h \cup h')(b) = \pi[(h \otimes h')D(b)]$.

Define $\varepsilon^*\colon \Lambda \to C^*(B;X)$ by $\varepsilon^*(1)(b) = \varepsilon(b)e$, e the identity of X, and define $\eta^*\colon C^*(B;X) \to \Lambda$ by $\eta^*(h) = h(\eta(1))$. Then $C^*(B;X)$ is a differential Λ-algebra with product $\cup$, unit ε^*, and augmentation η^*. The relation $\delta\cup = \cup\delta$ may be written out as:

(4) $\delta(h \cup h') = \delta(h) \cup h' + (-1)^{\deg h} h \cup \delta(h')$.

Together with Lemma 29.11, the constructions above imply

PROPOSITION 30.2: Let B be a simplicial set. Then $H_*(B)$ is a cocommutative coalgebra. If X is a differential Λ-module, then $H^*(B;H(X))$ is a Λ-algebra, which is commutative if $H(X)$ is commutative. (For the first statement, assume that $H_*(B)$ is Λ-flat.)

EXAMPLES 30.3: The most common instance of the proposition is $X = \Lambda$ with zero differential. We will have occasion to use the cases $X = C(G)$, where G is a simplicial group, and $X = C^*(F;\Lambda)$, where F is a simplicial set. In the latter case, we must take $X_{-n} = C^n(F;\Lambda)$ to make sense of the grading, and the cup product is here given explicitly (with π the product in Λ) by:

(5) $(h \cup h')(b)(f) = \pi[((h \otimes h')D(b))D(f)]$.

Next, let G be a simplicial group with product π, and suppose G operates from the left on a simplicial set F via $\sigma: G \times F \to F$. Let g denote the Eilenberg-MacLane map and define the Pontryagin product by:

(6) $\bar{\sigma} = C(\sigma)g: C(G) \otimes C(F) \to C(F)$.

With $F = G$ and $\sigma = \pi$, we find that $C(G)$ is a differential algebra with product $\bar{\pi}$, unit η (base point e_0) and augmentation ε. Then $C(F)$ is clearly a differential $C(G)$-module, and we have:

PROPOSITION 30.4: Let G be a simplicial group which operates from the left on a simplicial set F. Then $H_*(G)$ is a Hopf algebra, which is commutative if G is commutative, and $H_*(F)$ is a left $H_*(G)$-module. $H_*(F)$ is in fact a left module coalgebra over $H_*(G)$, in the sense that the module product is a morphism of coalgebras. (For both statements, assume that $H_*(G)$ is Λ-flat.)

Proof: We need only prove that the chain maps $D\bar{\sigma}$ and $(\bar{\sigma} \otimes \bar{\sigma})(1 \otimes T \otimes 1)(D \otimes D): C(G) \otimes C(F) \to C(F) \otimes C(F)$ are chain homotopic (taking $F = G$ and $\bar{\sigma} = \bar{\pi}$, this will imply that $H_*(G)$ is a Hopf algebra). Now if K and L are any simplicial Abelian groups, then by Theorem 28.3 (with $n = 1$, $s_0 = 0$, and the models of the previous section) the following diagram is chain homotopy commutative

$$\begin{array}{ccc} A(K) \otimes A(L) & \xrightarrow{g} & A(K \times L) \\ {\scriptstyle (1 \otimes T \otimes 1)(D \otimes D)} \downarrow & & \downarrow {\scriptstyle D} \\ A(K) \otimes A(L) \otimes A(K) \otimes A(L) & \xrightarrow{g \otimes g} & A(K \times L) \otimes A(K \times L) \end{array}$$

By the naturality of D, $D\sigma_* = (\sigma_* \otimes \sigma_*)D$. Therefore $D\sigma_* g$ and $(\sigma_* \otimes \sigma_*)(g \otimes g)(1 \otimes T \otimes 1)(D \otimes D)$ are chain homotopic, as was to be proven.

For the remainder of this section, B and F will denote simplicial sets and G will denote a simplicial group with product π which operates via σ on F. Define the cap product:

$$\cap: C^*(B;C(G)) \otimes [C(B) \otimes C(F)] \to C(B) \otimes C(F)$$

by

(7) $t \cap (b \otimes f) = (1 \otimes \bar{\sigma})(1 \otimes t \otimes 1)(D \otimes 1)(b \otimes f)$.

PROPOSITION 30.5: The cap product gives $C(B) \otimes C(F)$ a structure of left differential $C^*(B;C(G))$-module.

Proof: We must prove the following two formulae:

(8) $d(t \cap (b \otimes f)) = \delta(t) \cap (b \otimes f) + (-1)^{\deg t} t \cap d(b \otimes f)$,

(9) $(t \cup t') \cap (b \otimes f) = t \cap (t' \cap (b \otimes f))$.

Define $\mu(t)$: $C(B) \otimes C(F) \to C(B) \otimes C(G) \otimes C(F)$ by

$$\mu(t) = (1 \otimes t \otimes 1)(D \otimes 1),$$

$t \in C^n(B;C(G))$. To prove (8), it suffices to show that

$$d\mu(t) = \mu(\delta(t)) + (-1)^n \mu(t)d.$$

But:

$$\begin{aligned}
d\mu(t) &= (d \otimes 1 \otimes 1 + 1 \otimes d \otimes 1 + 1 \otimes 1 \otimes d)(1 \otimes t \otimes 1)(D \otimes 1)\\
&= [d \otimes t \otimes 1 + 1 \otimes dt \otimes 1 + (-1)^n(1 \otimes t \otimes d)](D \otimes 1)\\
&= [(-1)^n(1 \otimes t \otimes 1)(d \otimes 1 \otimes 1 + 1 \otimes d \otimes 1 + 1 \otimes 1 \otimes d) + 1 \otimes \delta(t) \otimes 1](D \otimes 1)\\
&= (-1)^n(1 \otimes t \otimes 1)(D \otimes 1)(d \otimes 1 + 1 \otimes d) + (1 \otimes \delta(t) \otimes 1)(D \otimes 1)\\
&= (-1)^n \mu(t)d + \mu(\delta(t)),
\end{aligned}$$

where we have used formula (2) and the relation $dD = Dd$. It remains to prove formula (9). Here we find:

$$\begin{aligned}
(t \cup t') \cap &= (1 \otimes \bar{\sigma})(1 \otimes \bar{\pi}(t \otimes t')D \otimes 1)(D \otimes 1)\\
&= (1 \otimes \bar{\sigma})(1 \otimes \bar{\pi} \otimes 1)(1 \otimes t \otimes t' \otimes 1)(1 \otimes D \otimes 1)(D \otimes 1)\\
&= (1 \otimes \bar{\sigma})(1 \otimes 1 \otimes \bar{\sigma})(1 \otimes t \otimes 1 \otimes 1)(1 \otimes 1 \otimes t' \otimes 1)(D \otimes 1 \otimes 1)(D \otimes 1)\\
&= (1 \otimes \bar{\sigma})(1 \otimes t \otimes 1)(1 \otimes 1 \otimes \bar{\sigma})(D \otimes 1 \otimes 1)(1 \otimes t' \otimes 1)(D \otimes 1)\\
&= (1 \otimes \bar{\sigma})(1 \otimes t \otimes 1)(D \otimes 1)(1 \otimes \bar{\sigma})(1 \otimes t' \otimes 1)(D \otimes 1)\\
&= t \cap (t' \cap), \text{ as desired.}
\end{aligned}$$

The cap product is natural in two senses, as shown in the following lemma.

LEMMA 30.6: Let $\beta\colon B \to B'$ be a simplicial map, let $\gamma\colon G \to G'$ be a simplicial homomorphism, and let $\alpha\colon F \to F'$ be a γ-equivariant map. If $t \in C^*(B;C(G))$ and $t' \in C^*(B';C(G'))$ satisfy $\gamma_*(t) = \beta^*(t')$, that is, if $C(\gamma) \circ t = t' \circ C(\beta)$, then:

(10) $(\beta_* \otimes \alpha_*)[t \cap (b \otimes f)] = t' \cap [\beta_*(b) \otimes \alpha_*(f)]$.

On the other hand, if $h \in C^*(B';C(G))$, then:

(11) $(1 \otimes \alpha_*)[h \cap (\beta_*(b) \otimes f)] = (\beta_* \otimes 1)[\gamma_*\beta^*(h) \cap (b \otimes \alpha_*(f))]$.

The proof is straightforward, and will be omitted.

REMARKS 30.7: If $F = G$ is the trivial simplicial group and we replace $C(G)$ by $C_N(G) = Z$, then the cap product reduces to a map $C^*(B) \otimes C_*(B) \to C_*(B)$. Explicitly, if $t \in C^p(B)$ and $b \in C_{p+q}(B)$, then:

(12) $t \cap b = (1 \otimes t)D(b) = (-1)^{pq}\tilde{\partial}^p b \cdot t(\partial_0^q b) \in C_q(B)$.

Now let Λ be a commutative ring. We can and will make the identification:

(13) $C^*(B;C^*(F;\Lambda)) = \mathrm{Hom}(C(B) \otimes C(F),\Lambda)$.

Then the dual of the cap product, which we shall also call a cup product, defines a structure of right differential $C^*(B;C(G))$-module on $C^*(B;C^*(F;\Lambda))$. Explicitly, if $h \in \mathrm{Hom}(C(B) \otimes C(F),\Lambda)$, we define:

(14) $(h \cup t)(b \otimes f) = h(t \cap (b \otimes f))$, and then

(15) $\delta(h \cup t) = \delta(h) \cup t + (-1)^{\deg h} h \cup \delta(t)$ and

(16) $h \cup (t \cup t') = (h \cup t) \cup t'$.

Here in deriving (15) we have used the special case of (2):

(17) $\delta(h)(b \otimes f) = (-1)^{\deg h+1} hd(b \otimes f)$.

At this point we have developed all the requisite machinery to define twisting cochains.

DEFINITION 30.8: Let $t \in C^1(B;C(G))$, so that $t_q: C_q(B) \to C_{q-1}(G)$. Define $d_t: C(B) \otimes C(F) \to C(B) \otimes C(F)$ by:

(18) $d_t(b \otimes f) = d(b \otimes f) + t \cap (b \otimes f)$.

Using (8) and (9), we find $d_t^2(b \otimes f) = (\delta(t) + t \cup t) \cap (b \otimes f)$. t is said to be a twisting cochain if $\delta(t) + t \cup t = 0$, that is, if

(19) $dt_n + t_{n-1}d + \sum_{i=1}^{n-1} t_i \cup t_{n-i} = 0, \quad n > 1,$

and if $\varepsilon t_1 = 0$ (so that $(\varepsilon \otimes \varepsilon)d_t = (\varepsilon \otimes \varepsilon)(t \cap) = 0$). Then d_t is called the differential twisted by t and $C(B) \otimes C(F)$ furnished with this differential is denoted by $C(B) \otimes_t C(F)$ and is called a twisted tensor product. Dually, $\mathrm{Hom}(C(B) \otimes_t C(F), \Lambda)$ is given the differential δ_t defined by (17) with δ_t and d_t replacing δ and d, or:

(20) $\delta_t(h) = \delta(h) + (-1)^{\deg h+1} h \cup t$.

§31. Brown's theorem

Brown's theorem states essentially that there is a natural way to assign to every twisting function τ a twisting cochain t in such a manner that $C(F \times_\tau B)$ is chain homotopy equivalent to $C(B) \otimes_t C(F)$. In the last section, this result will be used to construct the Serre spectral sequence.

Unless otherwise specified, the symbols C and C^* will denote the normalized chain and cochain functors in this section and the next, and the symbol (n) will refer to formula (n) of section 30.

We will use the method of acyclic models, and we must first define a category, the objects of which are all twisted Cartesian products.

DEFINITION 31.1: Let $F \times_\tau B$ and $F' \times_{\tau'} B'$ be TCP's with groups G and G', and let $\gamma: G \to G'$ be a simplicial homomorphism. A γ-map $\theta: E(\tau) \to E(\tau')$ is a simplicial map such that

$$\theta(f,b) = (\psi(b)\alpha(f),\beta(b)),$$

where $\alpha: F \to F'$ is a γ-equivariant map, $\beta: B \to B'$ is a simplicial map, and $\psi: B \to G'$ is a function. Clearly $p'\theta = \beta p$. We will write $\theta = (\alpha,\beta,\psi)$. θ is said to be γ-special if $\gamma\tau = \tau'\beta$. The requirement that θ be a simplicial map is equivalent to the identities:

$$\begin{array}{ll} & \tau'\beta(b)\partial_0\psi(b) = \psi(\partial_0 b)\gamma\tau(b) \\ (U) & \partial_i\psi(b) = \psi(\partial_i b) \text{ if } i > 0 \\ & s_i\psi(b) = \psi(s_i b) \text{ if } i \geq 0. \end{array}$$

The composite of a γ-map θ and a γ'-map θ' is the $\gamma'\gamma$-map $\theta'\theta = (\alpha'\alpha, \beta'\beta,(\psi'\beta)\cdot(\gamma'\psi))$. With the obvious identity maps, we have defined a category whose objects are all TCP's and whose maps are all γ-maps. If maps are required to be special, we obtain a subcategory with the same objects, which we shall call $\mathcal{R}$. $\mathcal{P}$ will denote the subcategory of $\mathcal{R}$, the objects of which are all PTCP's. Observe that if $\theta = (\alpha,\beta,\psi)$ is a γ-map of PTCP's, then necessarily $\alpha = \gamma$. If base complexes are required to be reduced, we obtain subcategories $\mathcal{R}_0$ of $\mathcal{R}$ and $\mathcal{P}_0$ of $\mathcal{P}$. The categories $\mathcal{R}_0$ and $\mathcal{P}_0$ will be of primary interest. (We implicitly make them additive; see § 28.)

The symbol $F \times_\tau B$ will denote ambiguously an object of $\mathcal{R}_0$ or the corresponding total complex. We now define model objects in the category $\mathcal{R}_0$. Let $\overline{\Delta}[n]$ denote $\Delta[n]/\Delta[n]^0$, where $\Delta[n]^0$ denotes the zero skeleton of $\Delta[q]$, and define the models $M^{p,q}$ of R_0 by:

(i) $M^{p,q} = (G(\overline{\Delta}[p] \times \Delta[q]) \times_{\tau(p)} \overline{\Delta}[p]$.

For clarity, we have here denoted the twisting function

$$\overline{\Delta}[p] \to G(\overline{\Delta}[p])$$

by $\tau(p)$. $G(\overline{\Delta}[p])$ operates on the fibre $G(\overline{\Delta}[p]) \times \Delta[q]$ via

$g(g',u) = (gg',u)$. Clearly

(ii) $M^{p,q} \cong (G(\overline{\Delta}[p]) \times_{\tau(p)} \overline{\Delta}[p]) \times \Delta[q]$.

Therefore the realization of each $M^{p,q}$ is contractible.

Using the models, we can assign a twisting cochain to each twisting function. We first need a definition.

DEFINITION 31.2: A twisting cochain on the category $\mathcal{P}_0$ is a function T which assigns a twisting cochain $T(\tau) \in \mathrm{Hom}^1(C(B),C(G))$ to each twisting function $\tau\colon B \to G$, B a reduced complex, in such a manner that the following conditions are satisfied:

(T.1) $T(\tau)(b) = e_0 - \tau^{-1}(b)$ for all nondegenerate $b \in B_1$

(T.2) If $\tau(b) = e_{q-1}$ for all $b \in B_q$ and all $q \leq n$, then $T(\tau)(b)=0$ for all nondegenerate $b \in B_q$ and all $q \leq n$.

(T.3) If $\theta = (\gamma,\beta,\psi)\colon G \times_\tau B \to G' \times_{\tau'} B'$ is a γ-special map of PTCP's, then the following diagram is commutative:

$$\begin{array}{ccc} C(B) & \xrightarrow{T(\tau)} & C(G) \\ \downarrow{\beta_*} & & \downarrow{\gamma_*} \\ C(B') & \xrightarrow{T(\tau')} & C(G') \end{array}$$

THEOREM 31.3: There exists a twisting cochain on the category $\mathcal{P}_0$.

Proof: Let $G \times_\tau B$ be a PTCP, B a reduced complex. Define $T(\tau)_1$ by formula (T.1) and define $T(\tau)_2$ by:

(T.4) $T(\tau)(b) = -\tau^{-1}(b) \cdot s_0\tau^{-1}(\partial_0 b)$ for all non-degenerate $b \in B_2$.

Clearly $\varepsilon \cdot T(\tau)_1 = 0$, and $dT(\tau)_2 + T(\tau)_1 d + T(\tau)_1 \cup T(\tau)_1 = 0$ is proven by an easy calculation. Condition (T.2) holds for $n = 2$ since e_1 is degenerate and therefore zero in $C(G)$. Suppose inductively that $T(\tau)_i$ has been defined for $i < q$, $q > 2$. We require that

$$dT(\tau)_q = -T(\tau)_{q-1}d - \sum_{i=1}^{q-1} T(\tau)_i \cup T(\tau)_{q-i} = X_q, \text{ say.}$$

Clearly $dX_q = 0$. First consider $G(\overline{\Delta}[q]) \times_{\tau(q)} \overline{\Delta}[q]$. Since $H_n(G(\overline{\Delta}[q])) = 0$ for $n > 0$, there exists $m \in C_{q-1}(G(\overline{\Delta}[q]))$ such that $d(m) = X_q(\Delta_q)$. We define $T(\tau(q))(\Delta_q) = m$. Next consider $G(K) \times_{\tau(K)} K$, $\tau(K)$: $K \to G(K)$, K any reduced complex, and let $x \in K_q$ be non-degenerate. $\overline{x}: \overline{\Delta}[q] \to K$ induces $G(\overline{x}): G(\overline{\Delta}[q]) \to G(K)$ and we define:

$T(\tau(K))(x) = G(\overline{x})_*(m) \in C_{q-1}(G(K))$. Then we have:

$$dT(\tau(K))(x) = G(\overline{x})_* d(m) = G(\overline{x})_* X_q(\Delta_q) = X_q(\overline{x}(\Delta_q)) = X_q(x).$$

Finally, consider the arbitrary PTCP $G \times_\tau B$ and let $b \in B_q$ be nondegenerate. If τ is induced by $f(\tau)$: $B \to \overline{W}(G)$, define:

$$T(\tau)(b) = \phi(f(\tau))_*[T(\tau(B))(b)] \in C_{q-1}(G), \text{ where}$$

$\phi(f(\tau))$: $G(B) \to G$ is as defined in Corollary 27.2. Then again we find $dT(\tau)(b) = X_q(b)$. Condition (T.2) holds since for any $c \in B_q$, $\phi(f(\tau))(\tau(B)(c)) = \tau(c)$, and since e_{q-1} is degenerate. Clearly condition (T.3) is satisfied, and this completes the proof.

We can now define the two functors A and B_T: $\mathcal{R}_0 \to \mathcal{C}$ that we wish to compare by the method of acyclic models. Thus define $A(F \times_\tau B) = C(F \times_\tau B)$ and $B_T(F \times_\tau B) = C(B) \otimes_t C(F)$, where $t = T(\tau)$, T being a fixed twisting cochain on the category $\mathcal{P}_0$. Observe that if $\theta = (\alpha, \beta, \psi)$ is a γ-special map of TCP's, then (T.3) and (10) of Lemma 30.6 guarantee that $B_T(\theta) = \beta_* \otimes \alpha_*$ is a chain map.

LEMMA 31.4: Both A_n and $(B_T)_n$ are representable, $n \geq 1$.

Proof: By Theorem 22.2, Corollary 22.3, and Lemma 29.5, it suffices to prove the result when $C(F \times_\tau B)$, $C(B)$, and $C(F)$ are interpreted as unnormalized chain complexes, $T(\tau)$ being extended to $C(B)$ by letting $T(\tau)(b) = 0$ if b is degenerate. Thus, for the purposes of this proof only, we regard A and B_T as being defined in terms of unnormalized chain complexes. Let $F \times_\tau B$ be a TCP with group G, B a reduced complex. If $b \in B_r$, define

$$\gamma(b) = \phi(f(\tau)) \circ G(\overline{b}): G(\Delta[r]) \to G,$$

where $f(\tau): B \to \overline{W}(G)$ induces τ and where $\phi(f(\tau))$ is as defined in Corollary 27.2. Explicitly, $\gamma(b)(\tau(r)(u)) = \tau \circ \overline{b}(u)$, $u \in \overline{\Delta}[r]$. Let $\sigma: G \times F \to F$ define the action of G on F and let $e: \overline{\Delta}[r] \to G$ be the map defined by $e(u) = e_m$, $u \in \overline{\Delta}[r]_m$. If $f \in F_s$, define $\theta(f,b): M^{r,s} \to F \times_\tau B$ to be the $\gamma(b)$-special map

$$\theta(f,b) = (\sigma(\gamma(b), \overline{f}), \overline{b}, e).$$

Now define $\xi: A_n \to \tilde{A}_n$ by

(i) $\xi(f,b) = (\theta(f,b), ((e_n, \Delta_n), \Delta_n))$, where $f \in F_n$, $b \in B_n$, and $((e_n, \Delta_n), \Delta_n) \in A(M^{n,n})$.

Then clearly $\lambda\xi = 1: A_n \to A_n$. Similarly, define $\xi: (B_T)_n \to (\tilde{B}_T)_n$ by

(ii) $\xi(b \otimes f) = (\theta(f,b), \Delta_p \otimes (e_q, \Delta_q))$, where $b \in B_p$, $f \in F_q$, $p+q = n$ and $\Delta_p \otimes (e_q, \Delta_q) \in B_T(M^{p,q})$.

Again, $\lambda\xi = 1: (B_T)_n \to (B_T)_n$, and this completes the proof.

LEMMA 31.5: $H_n(A(M^{p,q})) = 0$ and $H_n(B_T(M^{p,q})) = 0$, $n \geq 1$.

Proof: Since the realization of $M^{p,q}$ is contractible, $H_n(A(M^{p,q})) = 0$ is clear, $n \geq 1$. Now $\Delta[q]$ is contractible and the operation of $G(\overline{\Delta}[p])$ on $G(\overline{\Delta}[p]) \times \Delta[q]$ does not depend on $\Delta[q]$. It follows that:

(i) $B_T(M^{p,q})$ and $B_T(G(\overline{\Delta}[p]) \times_{\tau(p)} \overline{\Delta}[p])$ are chain homotopy equivalent.

(A more formal demonstration of (i) will be given in Remarks 31.6). Let $\pi = \pi_1(\overline{\Delta}[p])$. Then $H_q(G(\overline{\Delta}[p])) = 0$ if $q > 0$ and $H_0(G(\overline{\Delta}[p]))$ is isomorphic (under the Pontryagin product) to the group ring $Z(\pi)$. Now π has one generator x for each 1-simplex of $\Delta[p]$ and, regarding $Z(\pi)$ as a chain complex with differential zero, we may define a chain equivalence $h\colon CG(\overline{\Delta}[p]) \to Z(\pi)$ by $h(x) = 0$ if $\deg x > 0$ and $h(\tau(x)) = [x]$ if $x \in \overline{\Delta}[p]_1$. Since h is a morphism of algebras, we have:

(ii) $C(\overline{\Delta}[p]) \otimes_t CG(\overline{\Delta}[p])$ and $C(\overline{\Delta}[p]) \otimes_{h \circ t} Z(\pi)$ are chain homotopy equivalent, where $t = T(\tau(p))$.

(Again, a more formal argument will be given in Remarks 31.6). Now if $x \in \overline{\Delta}[p]_n$ is non-degenerate and $a \in \pi$, then

$$d_{ht}(x \otimes a) = \sum_{i=0}^{n} (-1)^i \partial_i x \otimes a + ht \cap (x \otimes a), \text{ and}$$

$$\begin{aligned} ht \cap (x \otimes a) &= (1 \otimes \sigma)(1 \otimes ht \otimes 1)(D \otimes 1)(x \otimes a) \\ &= \sum_{i=0}^{n} (1 \otimes \sigma)(1 \otimes ht \otimes 1)(\tilde{\partial}^{n-i} x \otimes \partial_0^i x \otimes a) \\ &= (-1)^{n-1} \partial_n x \otimes [t(\partial_0^{n-1} x)]a. \end{aligned}$$

Since $t(\partial_0^{n-1} x) = e_0 - \tau^{-1}(n)(\partial_0^{n-1} x)$ by (T.1), we find

$$\begin{aligned} d_{ht}(x \otimes a) &= \sum_{i=0}^{n} (-1)^i \partial_i x \otimes a + (-1)^{n-1} \partial_n x \otimes a + (-1)^n \partial_n x \otimes [\partial_0^{n-1} x]^{-1} a \\ &= \sum_{i=0}^{n-1} (-1)^i \partial_i x \otimes a + (-1)^n \partial_n x \otimes [\partial_0^{n-1} x]^{-1} a. \end{aligned}$$

By Definition 16.4, the last formula is precisely that for the boundary in the normalized chain complex of the universal covering complex of $\overline{\Delta}[p]$. Since all of the higher homology groups of this vanish, we have proven the result.

Before stating Brown's theorem, we define filtrations of $A(F \times_\tau B)$ and of $B_T(F \times_\tau B)$. Thus filter $C(F \times_\tau B)$ by:

(A) $(f,b) \in F^pC_n(F \times_\tau B)$ provided that $b = s_{j_q} \dots s_{j_1} b'$, where $0 \leq j_1 < \dots < j_q < n$, b' is non-degenerate, and $\dim b' = n - q \leq p$. $\Sigma z_i(f_i,b_i) \in F^pC_n(F \times_\tau B)$ provided that each $(f_i,b_i) \in F^pC_n(F \times_\tau B)$.

There results a spectral sequence $\{E^r(F \times_\tau B)\}$ which converges to $H_*(F \times_\tau B)$. Dualizing, let Λ be a commutative ring and filter $C^*(F \times_\tau B;\Lambda)$ by:

(A*) $F_pC^*(F \times_\tau B;\Lambda) = \mathrm{Hom}(C(F \times_\tau B)/F^{p-1}C(F \times_\tau B),\Lambda)$.

The resulting spectral sequence $\{E_r(F \times_\tau B)\}$ converges to $H^*(F \times_\tau B;\Lambda)$. The cup product preserves filtration, as is easily verified from the definition ((1) and (3)), and it follows that each E_r is a differential Λ-algebra. (See e.g. Massey [43, 44] for the construction of spectral sequences and of their products).

Filter $C(B) \otimes_t C(F)$, $t = T(\tau)$, by:

(B) $F^pC(B) \otimes_t C(F) = \sum_{i=0}^{p} C_i(B) \otimes C(F)$.

Observe that if $b \otimes f \in F^pC(B) \otimes_t C(F)$, then both $(d \otimes 1)(b \otimes f)$ and $t \cap (b \otimes f)$ are in $F^{p-1}C(B) \otimes_t C(F)$, the latter fact following from the definition (7) of $\cap$ and from the form of $D(b)$. Therefore $\bar{E}^1_{p,q}(F \times_\tau B) = C_p(B) \otimes H_q(F)$ (not necessarily as a complex) in the resulting spectral sequence $\{\bar{E}^r(F \times_\tau B)\}$. Dualizing, filter $\mathrm{Hom}(C(B) \otimes_t C(F),\Lambda)$ by:

(B*) $F_p\mathrm{Hom}(C(B) \otimes_t C(F),\Lambda) = \mathrm{Hom}(C(B) \otimes_t C(F)/F^{p-1}C(B) \otimes_t C(F),\Lambda)$.

Then $\bar{E}^{p,q}_1(F \times_\tau B) = C^p(B;H^q(F;\Lambda))$ in the resulting spectral sequence $\{\bar{E}_r(F \times_\tau B)\}$.

REMARKS 31.6: Suppose in the situation of (10) of Lemma 30.6 that a_* is a chain homotopy equivalence, $B = B'$, and $\beta = 1$, so that $1 \otimes a_*$ induces an isomorphism $\bar{E}^1(F \times_\tau B) \to \bar{E}^1(F' \times_{\tau'} B)$.

Then by the comparison theorem for spectral sequences (see e.g. Moore [5, 52]) it follows that $1 \otimes a_*: C(B) \otimes_t C(F) \to C(B) \otimes_{t'} C(F')$ is a chain homotopy equivalence. This argument applies to give a rigorous demonstration of (i) and (ii) in the proof of the lemma above.

We can now obtain Brown's theorem, which may be regarded as a direct generalization of the Eilenberg-Zilber theorem.

THEOREM 31.7: There exist natural filtration-preserving transformations $\phi: A \to B_T$ and $\psi: B_T \to A$ lying over the natural transformations $\phi_0 = T: A_0 \to (B_T)_0$ and $\psi_0 = T: (B_T)_0 \to A_0$ (where $T(x \otimes y) = y \otimes x$). Any two such ϕ are naturally chain homotopic via filtration preserving chain homotopies, as are any two such ψ. Further, for any such ϕ and ψ, $\phi \circ \psi: B_T \to B_T$ and $\psi \circ \phi: A \to A$ are naturally chain homotopic to the respective identity transformations via filtration-preserving chain homotopies.

Proof: To prove the existence of ϕ and ψ, we apply Theorem 28.2 with $n = 1$ (since $H_0(B_T(M^{p,q}))$ is non-trivial). Thus define:

$$\phi_1(f_1, b_1) = b_1 \otimes \tau(b_1)\, \partial_0 f_1 + \partial_1 b_1 \otimes f_1$$
$$\psi_1(b_0 \otimes f_1) = (f_1, s_0 b_0) \text{ and } \psi_1(b_1 \otimes f_0) = (s_0(\tau^{-1}(b_1) f_0), b_1).$$

Then an easy calculation using (T.1) proves that $\partial_t \phi_1 = \phi_0 \partial$ and $\partial \psi_1 = \psi_0 \partial_t$, and the existence of ϕ and ψ follows from Lemmas 31.4 and 31.5. The existence of the cited chain homotopies follows from Theorem 28.3 (with $n = 1$ and $s_0 = 0$). It remains to prove that ϕ, ψ, and the chain homotopies are filtration-preserving. Observe that the maps $\theta(f,b)$ defined in the proof of Lemma 31.4 satisfy

(i) $B_T(\theta(f,b))(B_T(M^{n,n})) \subset F^p C(B) \otimes_t C(F)$
if $(f,b) \in F^p C_n(F \times_\tau B)$, and

(ii) $A(\theta(f,b))(A(M^{q,r})) \subset F^pC(F \times_T B)$

if $b \otimes f \in F^pC_q(B) \otimes C_r(F)$, so that $q \leq p$.

Here (i) holds since $\overline{b}_*(u) = 0$ unless $m \leq p$, $u \in \overline{\Lambda}[n]_m$. (ii) holds since $\overline{b}_*(u)$ is here obtained from degeneracy operators applied to a nondegenerate element of dimension $\leq p$ if $u \in \overline{\Lambda}[q]_m$. Now an inspection of the proofs of Lemma 29.5 and Theorems 28.2 and 28.3 shows that ϕ, ψ, and the chain homotopies obtained above are all necessarily filtration-preserving.

§32. The Serre spectral sequence

Using Brown's theorem, we can easily develop the properties of the Serre spectral sequences $\{E^r(F \times_T B)\}$ and $\{E_r(F \times_T B)\}$, including the products in cohomology and the identification of the first non-trivial higher order differentials. We note first the following corollaries of Brown's theorem.

COROLLARY 32.1: $\phi: A \to B_T$ and $\psi: B_T \to A$ induce natural inverse isomorphisms $\phi^r: E^r \to \overline{E}^r$, $\psi^r: \overline{E}^r \to E^r$ and $\phi_r: \overline{E}_r \to E_r$, $\psi_r: E_r \to \overline{E}_r$, $r \geq 1$.

COROLLARY 32.2: Define a natural transformation:

(i) $D_T = (\phi \otimes \phi)D\psi: B_T \to B_T \otimes B_T$, where D is the natural coproduct $A \to A \otimes A$.

Then D_T is filtration-preserving and is chain homotopy coassociative and cocommutative. Dually, define

(ii) $\cup_T = \mathrm{Hom}(D_T, 1): \mathrm{Hom}(B_T, \Lambda) \otimes \mathrm{Hom}(B_T, \Lambda) \to \mathrm{Hom}(B_T, \Lambda)$.

Then each $\overline{E}_r$ is a differential Λ-algebra, and, for $r \geq 1$, ϕ_r and ψ_r are isomorphisms of differential Λ-algebras.

Now $\{E^r(F \times_T B)\}$, and therefore also $\{\overline{E}^r(F \times_T B)\}$, converges to $H_*(F \times_T B)$, and $\overline{E}^1_{p,q}(F \times_T B) = C_p(B) \otimes H_q(F)$. The dual statements hold for cohomology. To complete the development of the Serre spectral sequence, we must identify the terms $\bar{E}^2(F \times_\tau B)$ and $\bar{E}_2(F \times_\tau B)$. We need the following concepts.

DEFINITIONS 32.3: A twisting function τ is said to be n-trivial if $\tau(b) = e_{i-1}$ for all $b \in B_i$, $i \leq n$. A twisting cochain t is said to be n-trivial if $t_i = 0$, $i \leq n$. By condition (T.2), if τ is n-trivial, then so is $T(\tau)$.

REMARKS 32.4: If the group of the fibre bundle determined by $F \times_\tau B$ can be reduced to one such that $G_i = e_i$ for $i < n$, then $F \times_\tau B$ is isomorphic to $F \times_{\tau'} B$, where τ' is n-trivial. Conversely, if τ is n-trivial, then the subgroup G of $A(F)$ generated by $\{\tau(b)\}$ satisfies $G_i = e_i$ for $i < n$. If $\pi_i(B) = 0$ for $i \leq n$, then replacing $F \times_\tau B$ by a minimal fibre space of the same homotopy type (first applying $S \circ T$ to a minimal fibration of the same homotopy type, if necessary, to obtain Kan complexes), we may suppose that B_i has just one simplex, $i \leq n$, and then τ is clearly n-trivial.

Now let $F \times_\tau B$ be a TCP such that τ, and therefore $t = T(\tau)$, is n-trivial for some $n \geq 1$. By the definition (7) of $\cap$ and the form of $D(b)$, we find that if $b \otimes f \in F^pC(B) \otimes_t C(F)$, then $t \cap (b \otimes f) \in F^{p-n-1}C(B) \otimes_t C(F)$. It follows that $d^1 = d \otimes 1$ and $d^r = 0$, $2 \leq r \leq n$, in the spectral sequence $\bar{E}^r(F \times_\tau B)$. Therefore we have:

THEOREM 32.5: If τ is n-trivial for some $n \geq 1$, then

(i) $\bar{E}^{n+1}_{p,q}(F \times_\tau B) = \bar{E}^2_{p,q}(F \times_\tau B) = H_p(B; H_q(F))$, and

(ii) $\bar{E}^{p,q}_{n+1}(F \times_\tau B) = \bar{E}^{p,q}_2(F \times_\tau B) = H^p(B; H^q(F; \Lambda))$

Under the hypothesis of n-triviality, we can also identify

d^{n+1} and δ_{n+1}. Thus if $x \in \overline{E}^0(F \times_\tau B)$ survives to $\overline{E}^{n+1}(F \times_\tau B)$, then $d^{n+1}(x)$ is clearly induced by $t \cap x$ and if $h \in \overline{E}_0(F \times_\tau B)$ survives to $\overline{E}_{n+1}(F \times_\tau B)$, then $\delta_{n+1}(h)$ is induced by $(-1)^{\deg h+1} h \cup t$. By (19), since $t_i = 0$ for $i \leq n$, we have $dt_{n+1} = 0$ and

$$dt_{n+2} + t_{n+1}d = 0.$$

Thus $t_{n+1} \in \operatorname{Hom}(C_{n+1}(B), C_n(G))$ defines a cocycle

$$v \in \operatorname{Hom}(C_{n+1}(B), H_n(G)),$$

whose cohomology class in $H^{n+1}(B; H_n(G))$ we denote by ν. Now we have

(a) $\quad d^{n+1}(x) = \nu \cap x, \quad x \in H_*(B; H_*(F))$, and

(b) $\quad \delta_{n+1}(h) = (-1)^{\deg h+1} h \cup \nu, \quad h \in H^*(B; H^*(F; \Lambda))$

ν can be further identified as follows. Since we may assume that $G_i = e_i$ for $i < n$, we can define a cocycle $u \in C^n(G; H_n(G))$ by $u(g) = \{g\}$. u is called the fundamental cocycle of G and its cohomology class will be denoted by μ. Since B is reduced, say $B_0 = b_0$, we may consider u to be an element of

$$\operatorname{Hom}(C_0(B) \otimes_t C_n(G), H_n(G)),$$

and then $u \cup t$ is defined. Now if $b \in B_{n+1}$, then:

$$(u \cup t)(b \otimes e_0) = u(t \cap (b \otimes e_0)) = u(b_0 \otimes t(b)) = \{t(b)\} = v(b).$$

Thus $\mu \cup \nu = \nu$, hence $\delta_{n+1}(\mu) = (-1)^{n+1}\nu$ in $\overline{E}_{n+1}(G \times_\tau B)$ (with coefficient ring $H_*(G)$). In other words, ν is the transgression of $(-1)^{n+1}\mu$ in this spectral sequence. We have now proven

THEOREM 32.6: Let $F \times_\tau B$ be a TCP with group G such that $G_i = e_i$ for $i < n$. Let $\mu \in H^n(G; H_n(G)) = \overline{E}^{0,n}_{n+1}(G \times_\tau B)$ (with coefficients $H_*(G)$) denote the cohomology class of the fundamental cocycle of G. Let $\nu \in H^{n+1}(B; H_n(G)) = \overline{E}^{n+1,0}_{n+1}(G \times_\tau B)$ denote the transgression of $(-1)^{n+1}\mu$. Then d^{n+1} on $\overline{E}^{n+1}(F \times_\tau B)$ and δ_{n+1} on $\overline{E}_{n+1}(F \times_\tau B)$ (with coefficients Λ) are defined by formulae (a) and (b) above.

It remains only to identify the $\cup_T$-product induced on $\overline{E}_2(F \times_\tau B)$, where τ is 1-trivial. Now under the identification (13), the cup product (5) on $C^*(B; C^*(F; \Lambda))$ becomes:

$$\begin{aligned}\mathrm{Hom}(\overline{D},1)\colon \mathrm{Hom}(C(B) \otimes C(F), \Lambda) \otimes \mathrm{Hom}(C(B) \otimes C(F), \Lambda) \\ \to \mathrm{Hom}(C(B) \otimes C(F), \Lambda),\end{aligned}$$

where $\overline{D} = (1 \otimes T \otimes 1)(D \otimes D)$. Of course $D_T \neq \overline{D}$ in general. An easy acyclic model proof shows that $\overline{D}$ and D' are chain homotopic (on the category of Cartesian products), where $D' = (ft \otimes ft)DgT$. (Here D is the coproduct on $C(F \times B)$, and $t(f,b) = (b,f)$.) D' and D_T both define natural transformations $\overline{E}^1 \to \overline{E}^1 \otimes \overline{E}^1$ on the subcategory $\mathcal{O}$ of $\mathcal{R}_0$, the objects of which are all 1-trivial TCP's with reduced base complex. We have

THEOREM 32.7: D_T and $D'\colon \overline{E}^1 \to \overline{E}^1 \otimes \overline{E}^1$ are naturally chain homotopic on the category $\mathcal{O}$. Thus if τ is 1-trivial, then $\overline{E}_2(F \times_\tau B) = H^*(B; H^*(F; \Lambda))$ as a ring, where the product on the right is the usual cup product.

Proof: Take as models in $\mathcal{O}$ all induced TCP's $F \times_{\tau\overline{b}} \overline{\Delta}[p]$, where the reduced complex B is the base of a 1-trivial TCP $F \times_\tau B$ and $\overline{b}\colon \overline{\Delta}[p] \to B$. Then $\overline{E}^1_{p,*}$ is representable, $p \geq 1$, since if $\tilde{b}\colon F \times_{\tau\overline{b}} \overline{\Delta}[p] \to F \times_\tau B$ covers $\overline{b}$, then defining

$$\xi(b \otimes \{f\}) = (\tilde{b}, \Delta_p \otimes \{f\})$$

for $b \otimes \{f\} \in C_p(B) \otimes H_*(F)$ we find $\lambda\xi(b \otimes \{f\}) = b \otimes \{f\}$. We claim that if $p+q > 2$, then $\overline{E}^2_{p,*} \otimes \overline{E}^2_{q,*} = 0$ on the models of $\mathcal{O}$. To see this, we note first that $\pi = \pi_1(\overline{\Delta}[p])$ is a free group. In fact, π is freely generated by $\{(i, i+1) \mid 0 \leq i < p\}$. Next, if we choose a minimal subcomplex M of $ST(\overline{\Delta}[n])$, then M is a $K(\pi,1)$ and is therefore isomorphic to $\overline{W}(K(\pi,0))$ by Theorem 23.6. Since

$C(\overline{W}(K(\pi,0)))$ is isomorphic to the bar construction of the group ring $Z(\pi)$ (see e.g. MacLane [42, p. 114]), it follows that

$$H_*(\overline{\Delta}[n]) \cong H_*(\pi;Z)$$

(the cohomology of π with coefficients in Z, or $\mathrm{Tor}_*^{Z(\pi)}(Z,Z)$). Since π is free, $H_p(\overline{\Delta}[n]) = 0$ for $p > 1$ (see e.g. [42, p. 123]) and $H_1(\overline{\Delta}[n])$ is free Abelian. By the Kunneth theorem, this proves our claim. Now we will be able to obtain the desired chain homotopy by applying Theorem 28.3 with $n = 3$ and $s_0 = s_1 = s_2 = 0$ (using induction on the base degree in the proof), provided we can first prove that $D' = D_T$ on $\overline{E}^1_{p,*}$, $p = 0,1,2$. It is easily seen that we may suppose $\psi = gT$ on $F^1C(B) \otimes_t C(F)$ and $\phi = ft$ on $F^1C(F \times_\tau B)$ when we restrict ourselves to the category $\mathcal{O}$. Thus $D' = D_T$ on $\overline{E}^0_{0,*}$ and on $\overline{E}^0_{1,*}$. It remains to prove that D' and D_T are chain homotopic on $\overline{E}^0_{2,*}$. To prove this, let

$$\tilde{\Delta}[2] = \Delta[2]/\Delta[2]^1,$$

where $\Delta[2]^1$ denotes the 1-skeleton of $\Delta[2]$, and define new models in $\mathcal{O}$ by $M^{p} = (G(\tilde{\Delta}[2]) \times \Delta[p]) \times_{\tau(2)} \tilde{\Delta}[2]$. A proof similar to that of Lemma 31.4 proves that $\overline{E}^0_{2,p}$ is representable for $p \geq 1$. An argument analogous to that given above shows that if $p+q > 2$, then $\Sigma_{i+j=2} \overline{E}^1_{i,p} \otimes \overline{E}^1_{j,q} = 0$ on the models. Now we can apply Theorem 28.3 with $n = 3$ to obtain the desired chain homotopy, provided we first prove that $D' \simeq D_T$ on $E^0_{2,p}$, $p = 0,1,2$. Here a lengthy calculation gives the result, noting that on $\mathcal{O}$ we may suppose: $(b_0 = B_0)$

(i) $\phi(f_2,b_2) = b_2 \otimes \partial_0^2 f_2 + \partial_2 b_2 \otimes \tau(b_2)\partial_0 f_2 + b_0 \otimes f_2$
$\quad + b_0 \otimes s_0\tau^{-1}(b_2)s_1\tau(b_2)s_1\partial_0 f_2$

(ii) $\psi(b_2 \otimes f_n) = \sum_{(\mu,\nu)} (-1)^{\sigma(\mu)}(s_{\mu_n}\cdots s_{\mu_1} s_0\tau^{-1}(b_2)s_{\nu_2}s_{\nu_1}f_n,$
$s_{\mu_n}\cdots s_{\mu_1}b_2) \bmod F^1C(F \times_\tau B)$, where the sum is taken over all $(n,2)$-shuffles (μ,ν).

REMARKS 32.8: There are several comments to be made.

(1) Had we developed the theory of the operation of $\pi_1(B)$ on $\pi_n(F)$, then the assumption of 1-triviality on τ would have reduced to the assumption of n-simplicity for all n. However, the theory does not lend itself to the development of local coefficients.

(2) The treatment of products here is quite unsatisfactory. While it does show that products can be introduced in $\{\overline{E}_r(F \times_\tau B)\}$, it would be more desirable to have a procedure for defining a coproduct directly on $C(B) \otimes_t C(F)$, without having to throw the definition back to $C(F \times_\tau B)$. In case $C(B)$ is cocommutative, this can be done using a procedure developed in [45].

(3) The topological implications are clear. Starting with a Serre fibration $p: E \to B$, one can pass to a minimal sub-fibre space of the Kan fibration $S(p): S(E) \to S(B)$ and apply the theory above.

BIBLIOGRAPHICAL NOTES ON CHAPTER VI

The functor G was introduced by Kan [30, 31]. Our proof of Theorem 26.6 follows that of Cartan [6]. This construction generalizes results of Milnor [49], who defined the functors E, GE, and G^+E. The relationship between G and $\overline{W}$ was investigated by Kan [31]. Free simplicial groups have been studied by Kan in [34, 35, and 36]. Also Kan [29] has reproved the Hurewicz theorem by means of a study of the algebraic situation given by the map $p: G(K) \to A(K)$ and has shown in [36] that the exact sequence of Remarks 26.10 is essentially that of Whitehead [63].

The method of acyclic models was introduced by Eilenberg and MacLane in [12], and was used in [16] to prove the Eilenberg-Zilber theorem. Our treatment follows these sources and MacLane [42]. Most of the material of the last three sections is contained, in rather different form, in Gugenheim [19]. Section 30 relies heavily on suggestions of J. C. Moore. A systematic study of Hopf algebras may be found in Milnor and Moore [51]. The proof of Brown's theorem is parallel to, but simpler than, the topological proof given in his original paper [3]. An explicit expression for a twisting cochain $T(\tau)$ in terms of the twisting function τ has been obtained by Szczarba [61].

The Serre spectral sequence was, of course, studied in the classical paper [57], following its introduction in cohomology by Leray [38, 39]. The approach here shows that the introduction of cubical singular theory is unnecessary, a fact shown by Gugenheim and Moore [20] using quite different methods. Brown [3] proved that the spectral sequence defined here is in fact isomorphic to that defined by Serre.

Szczarba [61] studied the products in the Wang spectral sequence using twisted tensor products. The form of d_{n+1} and δ_{n+1} in the case of n-triviality was discovered by Fadell and Hurewicz [17], but of course the result is there proven by quite different methods. A generalization of this result is proven by Shih in [59].

BIBLIOGRAPHY

[1] M. G. BARRATT, V. K. A. M. GUGENHEIM, and J. C. MOORE. "On semisimplicial fibre bundles," *Am. J. Math.*, 81 (1959), 639–657.

[2] E. H. BROWN, JR. "Finite computability of Postnikov systems," *Ann. of Math.*, 65 (1957), 1–20.

[3] ____. "Twisted tensor products I," *Ann. of Math.*, 69 (1959), 223–246.

[4] H. CARTAN and J. P. SERRE. "Espaces fibres et groupes de homotopie, I," *C. R. Acad. Sci. Paris*, 234 (1952), 288–290.

[5] H. CARTAN *et al.* "Algèbres d'Eilenberg-MacLane et homotopie," *Séminaire Henri Cartan 1954–1955*, Paris, Ecole Normale Superieure, 1956.

[6] ____. "Quelques Questions de Topologies," *Séminaire Henri Cartan 1956–1957*, Paris, Ecole Normale Superieure, 1958.

[7] ____. "Invariant de Hopf et operations cohomologiques secondaires," *Séminaire Henri Cartan 1958–1959*, Paris, Ecole Normale Superieure, 1959.

[8] A. DOLD. "Homology of symmetric products and other functors of complexes," *Ann. of Math.*, 68 (1958), 54–80.

[9] A. DOLD and R. THOM. "Quasifaserungen und unendliche symmetrische Produkte," *Ann. of Math*, 67 (1958), 239–281.

[10] S. EILENBERG. "Singular homology theory," *Ann. of Math.*, 45 (1944), 63–89.

[11] S. EILENBERG and S. MACLANE. "Relations between homology and homotopy groups of spaces II," *Ann. of Math.*, 51 (1950), 514–533.

[12] S. EILENBERG and S. MACLANE. "Acyclic Models," *Am. J. Math.*, 75 (1953), 189–199.

[13] ____. "On the groups $H(\pi,n)$, I, II, and III," *Ann. of Math.*, 58 (1953), 55–106, 60 (1954), 49–139, 60 (1954), 513–557.

[14] S. EILENBERG and N. STEENROD. *Foundations of Algebraic Topology*, Princeton, Princeton University Press, 1952.

[15] S. EILENBERG and J. A. ZILBER. "Semi-simplicial complexes and singular homology," *Ann. of Math.*, 51 (1950), 499–513.

[16] ____. "On products of complexes," *Am. J. Math.*, 75 (1953), 200–204.

[17] E. FADELL and W. HUREWICZ. "On the structure of higher differential operators in spectral sequences," *Ann. of Math.* (2), 68 (1958), 314–347.

[18] V. K. A. M. GUGENHEIM. "On supercomplexes," *Trans. A.M.S.*, 85 (1957), 35–51.

[19] ____. "On a theorem of E. H. Brown," *Ill. J. Math.*, 4 (1960), 292–311.

[20] V. K. A. M. GUGENHEIM and J. C. MOORE. "Acyclic models and fibre spaces," *Trans. A.M.S.*, 85 (1957), 265–306.

[21] A. HELLER. "Homotopy resolutions of semi-simplicial complexes," *Trans. A.M.S.*, 80 (1955), 299–344.

[22] S. T. HU. *Homotopy Theory*, New York, Academic Press, 1959.

[23] I. M. JAMES. "Reduced product spaces," *Ann. of Math.*, 62 (1955), 170–197.

[24] D. W. KAHN. "Induced maps for Postnikov systems," *Trans. A.M.S.*, 107 (1963), 432–450.

[25] D. M. KAN. "Abstract Homotopy I, II," *Proc. Nat. Acad. Sci. USA*, 41 (1955), 1092–1096.

[26] D. M. KAN. "Adjoint functors," *Trans. A.M.S.*, 87 (1958), 294–329.

[27] ____. "Functors involving c.s.s. complexes," *Trans. A.M.S.*, 87 (1958), 330–346.

[28] ____. "On c.s.s. complexes," *Amer. J. Math.*, 79 (1957), 449–476.

[29] ____. "The Hurewicz theorem," *Proc. Inter. Sym. Algebraic Topology and Appl.*, Mexico, 1956.

[30] ____. "A combinatorial definition of homotopy groups," *Ann. of Math.*, 67 (1958), 282–312.

[31] ____. "On homotopy theory and c.s.s. groups," *Ann. of Math.*, 68 (1958), 38–53.

[32] ____. "On the homotopy relations for c.s.s. maps," *Bol. Soc. Mat.*, Mex., (1957), 75–81.

[33] ____. "On c.s.s. categories," *Bol. Soc. Mat.*, Mex., (1957), 82–94.

[34] ____. "Minimal free c.s.s. groups," *Ill. J. Math.*, 2 (1958), 537–547.

[35] ____. "Homotopy groups, commutators, and Γ-groups," *Ill. J. Math.*, 4 (1960), 1–8.

[36] ____. "A relation between CW-complexes and free c.s.s. groups," *Amer. J. Math.*, 81 (1959), 512–528.

[37] L. KRISTENSEN. "On secondary cohomology operations," *Math. Scand.*, 12 (1963), 57–82.

[38] J. LERAY. "L'anneau spectral et l'anneau d'homologie d'un espace localement compact et d'une application continue," *J. Math. Pures Appl.*, 29 (1950), 1–139.

[39] ____. "L'homologie d'un espace fibré dont la fibre est connexe," *J. Math. Pures Appl.*, 29 (1950), 169–213.

[40] S. MacLane. "Constructions simpliciales acycliques," *Colloque Henri Poincaré*, Paris, 1954.

[41] ____. *Simplicial Topology* (notes by Joseph Yao), Chicago Univ.

[42] ____. *Homology*, New York, Academic Press, 1963.

[43] W. S. Massey. "Exact couples in algebraic topology I–V," *Ann. of Math*, 56 (1952), 363–396, 57 (1953), 248–286.

[44] ____. "Products in exact couples," *Ann. of Math.*, 59 (1954), 558–569.

[45] J. P. May. "The cohomology of restricted Lie algebras and of Hopf algebras," J. of Algebra, 3(1966), 123–146.

[46] J. P. Meyer. "Principal fibrations," *Trans. A.M.S.*, 107 (1963), 177–185.

[47] J. Milnor. "Construction of universal bundles I, II," *Ann. of Math.*, 63 (1956), 272–284, 430–436.

[48] ____. "The geometric realization of a semi-simplicial complex," *Ann. of Math.*, 65 (1957), 357–362.

[49] ____. *The construction FK*, Princeton, 1956 (mimeographed notes).

[50] ____. "On spaces having the homotopy type of a CW-complex," *Trans. A.M.S.*, 90 (1959), 272–280.

[51] J. Milnor and J. C. Moore. "On the structure of Hopf algebras," *Ann. of Math.*, 81 (1965), 211–264.

[52] J. C. Moore. *Seminar on algebraic homotopy theory*, Princeton, 1956 (mimeographed notes).

[53] ____. *Semi-simplicial complexes and Postnikov systems.*

[54] M. M. Postnikov. Doklady Akad. Nauk SSSR (1951), 76, 3, 359–362; 76, 6, 789–791; 79, 4, 573–576.

[55] D. Puppe. "Homotopie und Homologie in Abelschen Gruppen und Monoidkomplexen I, II," *Math. Zeit.*, 68 (1958), 367–421.

[56] D. PUPPE. "A theorem on semi-simplicial monoid complexes," *Ann. of Math.*, 70 (1959), 379–394.

[57] J. P. SERRE. "Homologie singuliere des espaces fibrés," *Ann. of Math.*, 54 (1951), 425–505.

[58] ____. "Cohomologie modulo 2 des complexes d'Eilenberg-MacLane," *Comm. Math. Helv.*, 21 (1953), 198–232.

[59] W. SHIH. *Homologie des Espaces Fibrés*, Institut des Hautes Etudes Scientifiques, Publication Mathématiques no. 13 (1962).

[60] N. E. STEENROD and D. B. A. EPSTEIN. "Cohomology Operations," *Ann. of Math. Studies* 50, Princeton, 1962.

[61] R. H. SZCZARBA. "The homology of twisted Cartesian products," *Trans. A.M.S.*, 100 (1961), 197–216.

[62] J. H. C. WHITEHEAD. "Combinatorial homotopy I," *Bull. A.M.S.*, 55 (1959), 213–245.

[63] ____. "A certain exact sequence," *Ann. of Math.*, 52 (1950), 51–110.